AF358646

Advances in Pyrometallurgy

Advances in Pyrometallurgy

Editor

Ari Jokilaakso

MDPI • Basel • Beijing • Wuhan • Barcelona • Belgrade • Manchester • Tokyo • Cluj • Tianjin

Editor
Ari Jokilaakso
Chemical and Metallurgical
Engineering
Aalto University
Espoo
Finland

Editorial Office
MDPI
St. Alban-Anlage 66
4052 Basel, Switzerland

This is a reprint of articles from the Special Issue published online in the open access journal *Metals* (ISSN 2075-4701) (available at: https://www.mdpi.com/journal/metals/special_issues/advances_pyrometallurgy).

For citation purposes, cite each article independently as indicated on the article page online and as indicated below:

LastName, A.A.; LastName, B.B.; LastName, C.C. Article Title. *Journal Name* **Year**, *Volume Number*, Page Range.

ISBN 978-3-0365-1849-7 (Hbk)
ISBN 978-3-0365-1850-3 (PDF)

Contents

About the Editor

Ari Jokilaakso

Ari Jokilaakso, Associate Professor (tenured) Aalto University, School of Chemical Engineering, Department of Chemical and Metallurgical Engineering. D.Sc. in 1992 Helsinki University of Technology. Professor Jokilaakso (M-7223-2016) has started in September 2016 as professor and leader of the Research Group for Metallurgy at Aalto University. He has a solid experience in experimental and computer simulation research of high temperature metallurgical processes. Prof. Jokilaakso has an exceptional career combining scientific basic research and industrial experience from research, engineering and technology management, and global leadership positions, including leader of the research center (200 employees), and being a member of the Executive Board of a stock listed company. Currently, he is also Vice Head, Department of Chemical and Metallurgical Engineering, Aalto University, School of Chemical Engineering, and Guest Professor (April 2018–April 2023) at Central South University, Changsha, P.R. of China.

Prof. Jokilaakso is currently (2021) involved in many circular economy of metals related academic and industrial research projects dealing with,but not limited to, non-ferrous slag reduction cleaning with alternative reductants, recovery of battery metals from End-of-Life battery scrap by using existing pyrometallurgical processes, thermophysical properties measurement of slags, mattes and refractories, and carbon avoidance in steel and ferroalloys production. He has authored and coauthored more than 220 scientific articles, conference presentations, reports and other publications. Has has been recognized with three best paper awards: Billiton Medal 1999 The best papers published 1998 in the Transactions of The Institution of Mining and Metallurgy, London, UK, for the papers entitled:

- S.K. Strömberg, A.T. Jokilaakso, S.K. Jyrkönen: Oxidation behaviour of violarite-based nickel concentrate in simulated suspension smelting conditions. Trans. Inst.Min.Metall. (Sec. C: Mineral Process. Extr. Metall.), 107, January–April 1998, C18–C29.

- S.K. Jyrkönen, A.T. Jokilaakso, S.K. Strömberg, P.A. Taskinen: Behaviour of synthetic nickel mattes under suspension smelting conditions. Trans. Inst.Min.Metall. (Sec. C: Mineral Process. Extr. Metall.), 107, January–April 1998, C30–C36.

The Gold Fields of South Africa, Ltd, Gold Medal and Premium The best paper published 1991 in the Transactions of The Institution of Mining and Metallurgy, London, UK, for the paper entitled

- Jokilaakso, R. Suominen, P. Taskinen, K. Lilius: Oxidation of chalcopyrite in simulated suspension smelting. Trans. Inst.Min.Metall. (Sec. C: Mineral Process. Extr. Metall.), 100, May–Aug. 1991, C79–C90.

Article

Experimental Model Study of Liquid–Liquid and Liquid–Gas Interfaces during Blast Furnace Hearth Drainage

Weiqiang Liu [1,*], Lei Shao [2] and Henrik Saxén [1]

[1]   Process and Systems Engineering Lab., Åbo Akademi University, Åbo 20500, Finland; henrik.saxen@abo.fi
[2]   School of Metallurgy, Northeastern University, Shenyang 110819, China; shaolei@smm.neu.edu.cn
*   Correspondence: weiqiang.liu@abo.fi; Tel.: +358-449770584

Received: 3 March 2020; Accepted: 7 April 2020; Published: 9 April 2020

Abstract: The smooth drainage of produced iron and slag is a prerequisite for stable and efficient blast furnace operation. For this it is essential to understand the drainage behavior and the evolution of the liquid levels in the hearth. A two-dimensional Hele–Shaw model was used to study the liquid–liquid and liquid–gas interfaces experimentally and to clarify the effect of the initial amount of iron and slag, slag viscosity, and blast pressure on the drainage behavior. In accordance with the findings of other investigators, the gas breakthrough time increased and residual ratios for both liquids decreased with an increase of the initial levels of iron and slag, a decrease in blast pressure, and an increase in slag viscosity. The conditions under which the slag–iron interface in the end state was at the taphole and not below it were finally studied and reported.

Keywords: blast furnace; hearth drainage; iron and slag flow; interface phenomena

---

1. Introduction

The ironmaking blast furnace (BF), which will likely remain the dominant process in supplying hot metal for the production of crude steel in the near future, has undergone remarkable developments in both operating efficiency and working volume. The growing size of the blast furnace has made it more difficult to maintain a uniform distribution over the cross section, which can lead to permeability problems in the coke bed (dead man) in the lower part of the process [1]. A higher flow resistance for the hearth liquids may cause hearth drainage problems, which are intimately associated with excessive/abnormal wear of hearth lining and upsets of the operation state. Therefore, smooth drainage is of crucial importance to the hearth integrity that is today commonly recognized as the main factor limiting BF campaign life.

For a well-controlled drainage of the hearth, both the slag–gas interface (henceforth called the "l–g interface") and the slag–iron interface (henceforth referred to as the "l–l interface") should stay at relatively low vertical levels in the hearth. Sometimes, however, incomplete drainage may occur, often caused by problems in extracting slag of high viscosity or at high draining rates. In such situations, the resulting l–g interface in the hearth may rise excessively, which has an adverse effect on the BF operation, resulting in, e.g., unstable burden descent. In the most extreme situation, the l–g interface can even rise up to the level of the tuyeres, causing serious problems in the blast supply, or in the worst case explosion due to water leakage, which can endanger the safety of the casthouse personnel [2].

During most of the draining, iron and slag flow out simultaneously, but single-phase flow may occur if the entrance of the taphole is immersed in only one of the liquid phases at the start of drainage. This results in periods of slag-only or iron-only tapping. The former case, where slag is the only phase to flow out initially, is often accompanied by a rapid upward bending of the l–l interface towards the taphole [3,4]. As the l–l interface has reached the taphole, both liquids flow out simultaneously until the end of the drainage. In the latter case, the l–l interface is (clearly) above the taphole so iron is the first phase to drain. During this period, both interfaces remain essentially horizontal, except for a slight declivity of the l–l interface close to the taphole when the overall l–l interface has descended sufficiently. As the local l–l interface reaches the taphole, both phases drain simultaneously, and the l–l interface descends below the taphole. Thus, both interfaces tilt toward the taphole until the draining period ends by the escape of gas through the taphole [5]. For most of the drainage situations, the local l–g interface tilts downwards to the taphole and the local l–l interface tilts upwards, as schematically illustrated in Figure 1. These complex phenomena are the result of different densities and viscosities of the two phases: a substantial local pressure gradient is formed in front of the taphole as the highly viscous slag flows through the dead man to the taphole, and there is accelerated flow in the vicinity of the taphole [2–7]. The velocity difference between the interior of the hearth and the region close to the taphole is caused by the large dimensional difference between the diameters of the BF hearth and the taphole. Furthermore, some special interface phenomena may occur in the hearth during drainage. For drainage of a hearth with a heterogeneous dead man, so-called viscous fingering may occur in the l–g interface close to the taphole [8,9]. Furthermore, for large BFs, it is estimated that the vertical level of the interfaces may vary in different parts of the hearth as a result of an impermeable dead man [10–12]. Such special phenomena can make the motion of the interfaces far more complicated than the expected overall behavior outlined above.

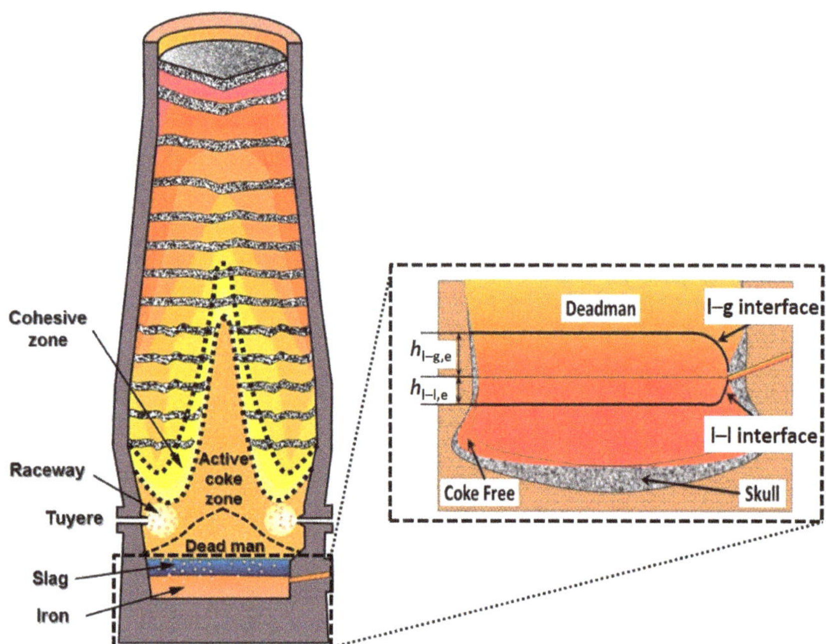

Figure 1. Schematic vertical cross-section of the blast furnace. l–g interface: slag–gas interface; l–l interface: slag–iron interface; $h_{l-l,e}$: final level of the horizontal part of the l–l interface; $h_{l-g,e}$: final level of the horizontal part of the l–g interface.

Due to the importance and complexity of the behavior of the interfaces, many studies of the interface phenomena in the hearth have been done in small-scale experimental models and using sophisticated numerical simulation models. Tanzil et al. [3–5] were the first to shed light on the general motion and bending of the interfaces in the BF hearth during drainage, partly revising the findings of Fukutake and Okabe [13]. Detailed investigations of the interface phenomena have been conducted based on the general findings by Tanzil et al. [3–5] and by Zulli [6]. Researchers have established simplified mathematical models estimating the l–l and l–g interface levels in the BF hearth as offline tools [10–12,14] or online based on measurement data [15,16]. Efforts have also been made to study the sophisticated interface phenomena by Computational Fluid Dynamics (CFD) [17] or a combination of this technique with the Discrete Element Method (CFD–DEM) [18–20]. Even though there are many merits of numerical simulation, e.g., low economic cost, high efficiency, etc., experimental studies are still critically essential for verification of numerical models and for studying certain phenomena for which computational analysis is still cumbersome. Experiments studying the effect of operation conditions, such as coke-free zones, hearth coke permeability, etc., on the behavior of the interfaces in batch or continuously operated small-scale physical models have been reported [3–6,8,9,11]. Compared with simulation studies, the experimental investigations are far more laborious, both when it comes to undertaking the experiments and when analyzing and compiling the results to yield generic findings. Still, the experimental approach is sometimes justified as it studies the "real" system, albeit in a simplified form.

In order to gain a better view of the complex evolution of the liquid levels in the blast furnace hearth, a series of experiments has been conducted in a two-dimensional Hele–Shaw (H–S) model; this kind of viscous flow analog model of the conditions in the BF hearth was used extensively in the pioneering studies in Australia [3–6,8]. The present work was primarily aimed at a systematical investigation of the effect of operation parameters such as the initial amount of iron and slag, slag viscosity, and blast pressure on the gas breakthrough time, as well as the interface states at the termination of drainage. In practice, the permeability of the packed bed also affects the drainage process, but this factor was constant here because the spacing between the plates in the pilot model was fixed. Pictures taken by a high-speed camera during the experiments were interpreted by an automatic interface tracking program to determine both the l–l and l–g interfaces accurately during the draining process, and the results will be described in this paper. The section that follows describes the apparatus arrangement and procedure as well as experimental conditions and analytical methods. Section 3 presents the results of the experiments that are analyzed and illustrated in both dimensional and dimensionless forms. Finally, in Section 4 some conclusions are drawn on the basis of the results of the study.

2. Experimental Setup

2.1. Apparatus Arrangement and Procedure

The experimental apparatus, as shown in Figure 2, consisted of three main subsystems, i.e., the draining system, charging system, and recording system. The draining system included an H–S model and a liquid receiver. The H–S model consisted of two parallel Perspex plates (585-mm wide and 580-mm high) which were kept at a uniform distance of 2 mm by Perspex strips along the bottom and sidewalls and a number of mini-spacers inside of the model. Three ports were cut in the two sides of the H–S model to allow for fluid supply or extraction. One of the ports, the outlet, was connected by a hose to a liquid receiver that was held at vacuum pressure by a pump to which it is connected. The volume of the receiver was large enough to keep the vacuum pressure practically constant during the draining process, so the pressure was set prior to the experiment. Oil and water were used as the fluids in the study, while the upper part of the model was open, so atmospheric pressure prevailed at the upper oil surface in the experiments. The oil supply system consisted of a moveable oil distributor, an oil reservoir, and a variable-speed peristaltic pump. The moveable oil distributor included of a rectangular mini-container connected to a group of tiny tubes. The container was connected to an oil reservoir by a hose with a variable-speed peristaltic pump. During the oil charging process, the tiny tubes were introduced into the slot of the H–S model, reaching down to the +120 mm level with reference to the outlet. This arrangement was necessary to be able to create a uniform and horizontal oil layer in a relatively short time, after which the pipes were withdrawn and the mini-container was removed (so as not to interfere with the images). The water charging system was very simple since the water supply could be controlled easily: a tube was connected from a water reservoir to the water inlet of the model and the flow rate was adjusted manually using a ball valve. A lifting table was used to support the oil and water reservoirs at a relatively high level for injecting water and oil into the model by gravity. The water supply rate was controlled by adjusting the height of the lifting table and the opening of the ball valve. The supplementary system (i.e., recording system) was a combination of a high-speed camera and laptop, which was used to record the whole drainage process of the experiment. By using the high-speed camera, the starting and terminal points of the drainage process could be determined accurately and more details about the drainage phenomena could be efficiently captured.

As the core part of the apparatus, the H–S viscous flow model emulates the two-phase flow in the BF hearth (i.e., the flow of molten iron and slag) by using mineral oil and water in the model. The approach is based on a flow analogue between viscous liquid flow between two closely parallel plates and flow in a packed bed [4]. Using the H–S model instead of a packed bed model solves the flow visualization problem. The physical properties of the above fluids are reported in Table 1.

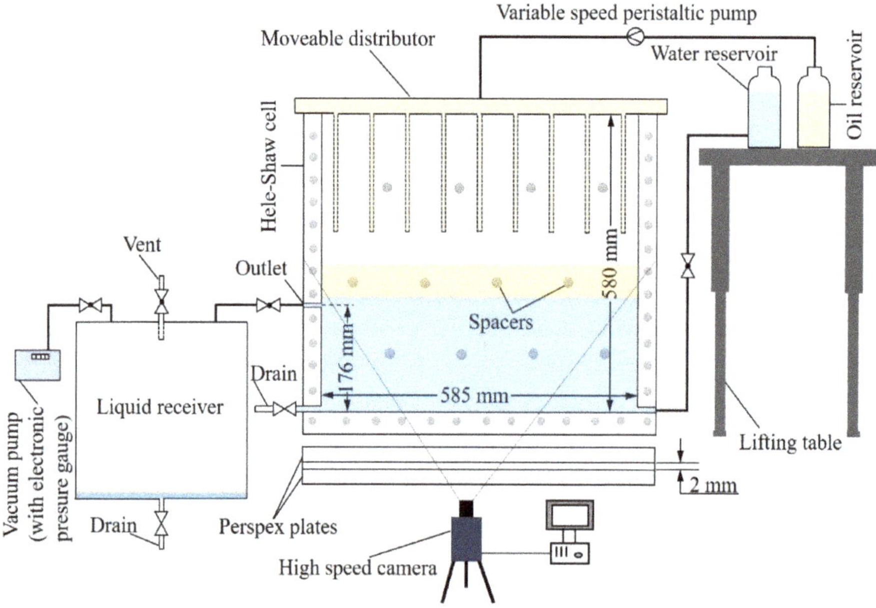

Figure 2. Schematic illustration of the experimental setup.

Table 1. Physical properties of fluids for the blast furnace (BF) and model system.

Properties	System	
	BF: Slag–Iron	Model: Oil–Water
Density of phase 1, $\rho_{L,1}$ (kg/m^3)	6800	998
Dynamic viscosity of phase 1, $\mu_{L,1}$ (Pa·s)	0.0068	0.001
Density of phase 2, $\rho_{L,2}$ (kg/m^3)	2800	855
Dynamic viscosity of phase 2, $\mu_{L,2}$ (Pa·s)	0.43	0.131, 0.254

According to the viscous flow analogue, Darcy flow in a packed bed with hydraulic conductivity, K, can be described with the H–S model (as explained in detail in [4]), if

$$K = \frac{\rho_L g b^2}{12 \mu_L} \tag{1}$$

where b is the spacing between the plates in the H–S model, g is the gravitational acceleration, μ_L is the liquid viscosity, and ρ_L is the liquid density.

The definition of hydraulic conductivity in a packed bed model is

$$K = \frac{k \rho g}{\mu} \tag{2}$$

where k is the absolute permeability of the packed bed. Combining the two equations above, we obtain

$$\left(\frac{k \rho g}{\mu} \right)_{Packed-bed} = \left(\frac{\rho g b^2}{12 \mu} \right)_{H-S} \tag{3}$$

In the experiments, the oil and water flow through the H–S model emulates iron and slag flow in the BF hearth, so

$$\frac{k \rho_{iron} g}{\mu_{iron}} = \frac{\rho_{water} g b^2}{12 \mu_{water}} \tag{4}$$

$$\frac{k \rho_{slag} g}{\mu_{slag}} = \frac{\rho_{oil} g b^2}{12 \mu_{oil}} \tag{5}$$

Since the fluids in the BF hearth and in the experimental model show practically equal density-to-viscosity ratios, a relation between the plate spacing, b, and the absolute permeability of the packed bed is obtained as

$$b = \sqrt{12k} \tag{6}$$

Thus, by adjusting the space of two plates in the H–S model, it is possible to simulate the iron and slag flow in the dead man with a wide range of absolute permeability for the H–S model (although the ratio of hydraulic conductivities of the two liquids remains unchanged). In the BF hearth, a typical effective coke particle diameter is 35 mm and the porosity is 0.30–0.40. According to the empirical relationship between effective particle diameter and absolute permeability [21], a rough estimate of the permeability is $k \approx 10^{-7}$ m^2. By choosing $b = 2$ mm, the absolute permeability of the BF hearth coke bed can be simulated by the H–S model.

The experimental procedure includes the following main steps:

(1) The lifting table was elevated to a certain level, followed by opening the ball valve to charge water into the model until the predetermined level was reached.

(2) The moveable oil distributor was fixed on top of the model and the peristaltic pump was started at low pumping speed to feed oil into the model slowly to create an oil layer with uniform and desired thickness.

(3) A settling time of a few minutes was allowed for both l–g and l–l interfaces to become absolutely stable. After this, the oil distributor was removed from the top of the model.

(4) The vacuum pump was switched on until the desired under-pressure was obtained in the receiver.

(5) The high-speed camera was turned on to record the drainage process.

(6) The outlet was opened fully. When air started blowing out through the outlet, the outlet was closed and the camera was switched off.

(7) The vent was opened to recover the receiver pressure, and the drain of the receiver was opened to empty the receiver.

2.2. Experimental Conditions and Analytical Methods

For the industrial BF, before the start of tapping, the blast pressure, l–l and l–g interface levels, and slag viscosity may vary from tap to tap, and all of these conditions influence the drainage of the hearth. Thus, it is interesting to study the influence of the above conditions on the tapping in the experimental model to gain insight into the practical drainage behavior of the BF hearth.

In the experiments, four conditions, i.e., pressure difference, initial l–l interface level, oil viscosity, and initial oil layer thickness, were chosen to investigate the drainage phenomena. The experimental conditions are listed in Table 2. The first group of experiments was the benchmark for the full set of experiments. In the table, the pressure difference is the difference between atmospheric pressure (that acts on the l–g interface) and the pressure inside the receiver. The initial l–l interface level, $h_{l-l,0}$, is expressed with respect to the level of the outlet. In the BF, it depends on the accumulated iron amount in the hearth, and on the length and outer level of taphole, since the true taphole tilts downwards. In the H–S model, it is not convenient to study the effect of the taphole length, so the initial l–l level was directly affected by the amount of water fed into the model. The reported initial oil layer thickness, $h_{oil,0}$, is the absolute thickness in the model.

After each group of experiments, the videos recorded during the drainage process were processed to extract the required information. Since the experimental model was two-dimensional, it was feasible to estimate the remaining volume of oil and water in the model by calculating the oil and water areas in every picture and multiplying them by the spacing between the two plates. A routine was written in MATLAB R2018a (by The MathWorks, MA, USA) to automatically detect the l–l and l–g interfaces to determine the interface locations. Then the areas of both liquids and finally the residual volume could be easily determined for every frame of the video. The method for automatic interface detection and processing is described elsewhere [22]. Figure 3 illustrates schematically the initial and end states of a hypothetical case.

Table 2. Experimental conditions. Changed parameters are written in bold. The dynamic viscosity of water is 0.001 Pa·s for all the experiments.

Experimental Group Number	Δp bar	Initial l–l Interface Level mm	Oil Viscosity Pa·s	Initial Oil Layer Thickness mm
1	**0.3**	**10**	**0.131**	**20–100**
2	**0.2**	10	0.131	-
3	**0.4**	10	0.131	-
4	0.3	**5**	0.131	-
5	0.3	**20**	0.131	-
6	0.3	10	**0.254**	-

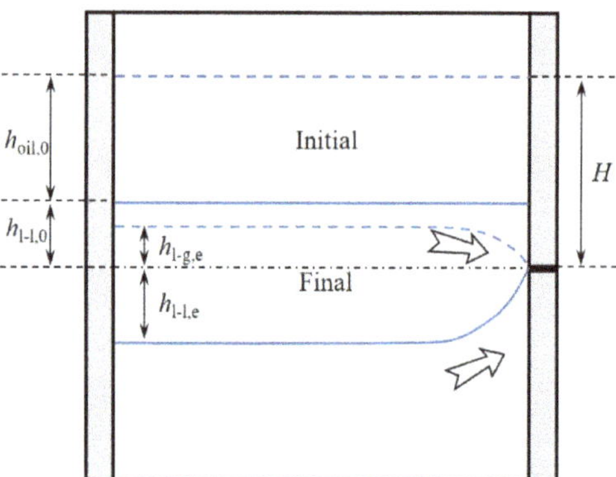

Figure 3. Schematic illustration of initial and final interface states. Solid lines: l–l interface, Dashed lines: l–g interface.

To use the experimental results for a real BF, some results were dealt with in a dimensionless format, i.e., the residual oil (or water) ratio (α), dimensionless gas breakthrough time (τ), and flow-out coefficient (F_L), where the last factor is an important dimensionless variable characterizing the drainage [6]. The definitions of the above parameters are

$$\alpha = \frac{V_{\text{end}}}{V_{\text{start}}} \tag{7}$$

$$\tau = \frac{t}{t_{\text{ave}}} = \frac{Qt}{Db(h_{\text{oil},0} + h_{\text{l}-\text{l},0})} \tag{8}$$

$$F_L = \left\{ 180 \frac{(1-\varepsilon)^2}{\varepsilon^3} \cdot \frac{1}{\varphi^2 d^2} \cdot \frac{\mu}{\rho g} \right\} U_0 \left[\frac{D}{H} \right]^{1.94} = \frac{\mu}{k\rho g} \cdot U_0 \cdot \left[\frac{D}{H} \right]^{1.94} \tag{9}$$

where V_{start} and V_{end} are the total oil or water volume at the start and end of drainage, respectively. In the equations, D is the width of the H–S model, t is the gas breakthrough time, and t_{ave} is the time taken to drain the whole liquids above the outlet at the average tapping rate, Q. Furthermore, $h_{\text{l}-\text{l},0}$ is the initial level of the l–l interface above the outlet, $h_{\text{oil},0}$ is the initial absolute oil layer thickness in the model, ϕ is the shape factor of coke particles, ε is the bed porosity, d is the particle diameter, μ is the viscosity and ρ the density of oil, U_0 is the superficial velocity, k is the absolute permeability of coke bed in hearth, and H is the initial level of l–g interface above the outlet.

3. Results and Discussion

In order to provide a general view of the evolution of the interfaces, Figure 4 shows an example of the liquid levels detected at five different moments during the process of a draining experiment. The figure shows how the l–g interface (blue line) from being horizontal gradually bends downwards towards the outlet along with the progress of the draining, while, by contrast, the l–l interface starts bending upwards after the interface in question has reached the level of the outlet. It is also seen that the l–g interface bends more than the l–l interface, which is due to differences in viscosity and density of the two liquid phases.

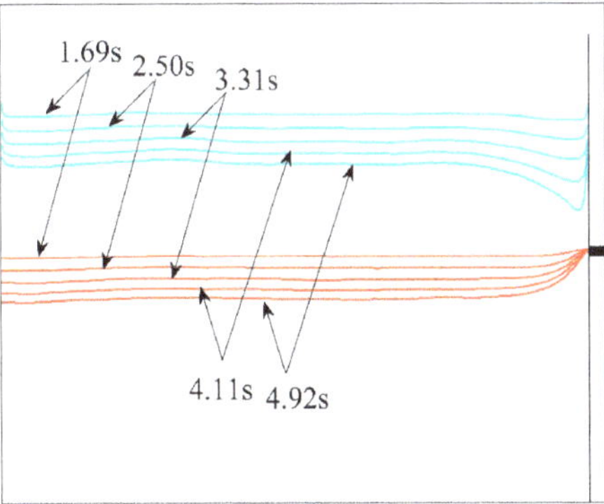

Figure 4. Interfaces between oil and gas (l–g, blue lines) and water and oil (l–l, red lines) at five different moments during a draining experiment.

3.1. Influence on Residual Ratio, Tapping Time

For the BF operation, the residual ratio of both liquids and tapping time are critical parameters. Under certain conditions, a longer tapping time may mean a higher operation rate and lower economic costs, because with extended tapping the residual liquid ratios and frequency of drilling and plugging of the taphole decrease, thereby reducing the cost of refractory materials and taphole clay.

3.1.1. Influence of Initial Oil Layer Thickness

In the BF, the amount of molten slag depends on the grade of iron ore, quality of coke, and the amount of injected pulverized coal, so it may fluctuate with the quality and quantity of raw materials used. If the slag ratio (i.e., the ratio of produced slag and hot metal) is high, the accumulated slag layer is thicker and there is more slag in the BF hearth during the drainage. To study the influence of the thickness of the accumulated slag layer on the tapping, a group of experiments was conducted, with results shown in Figure 5, where the experiments acted as benchmarks.

The upper panel of the figure shows that, as expected, the draining time increased with the initial oil layer thickness. This is simply because the distance between outlet and l–g interface increased since the initial l–l interface level was kept constant. The lower panel of the figure illustrates that both residual liquid ratios decreased with the initial oil layer thickness and that the decrease of the residual water ratio was more significant. The primary reason for the lower water residual ratio was that the thicker initial oil layer delayed the moment when the oil surface bent downwards to the outlet, which was the end point of the drainage, and the l–l interface thus had longer time to descend. This is also the reason why the residual oil ratio was smaller, as illustrated in the schematic in Figure 6; even though the water level is lower, the residual oil ratio decreased as the end l–g level increased only slightly.

Thus, for the BF process, with less produced slag, the draining time would be shorter. As this would lead to more frequent taps, implying higher labor and refractory costs, it is necessary to decrease the taphole diameter to extend the draining time in such situations.

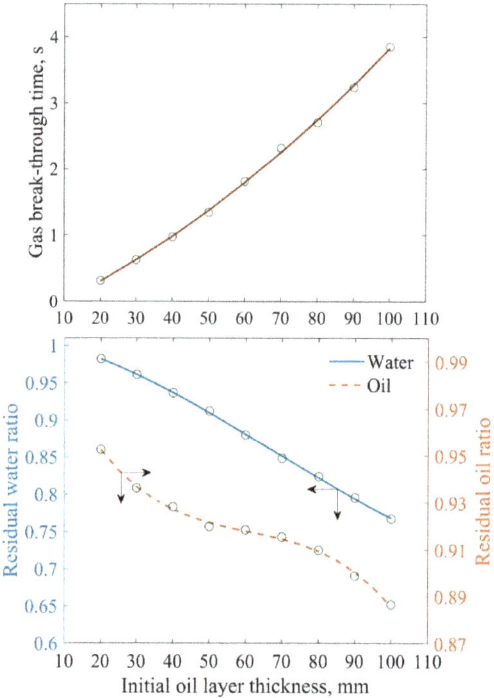

Figure 5. Upper panel: Effect of initial oil layer thickness on gas breakthrough time. Lower panel: Effect of initial oil layer thickness on residual ratios of both liquids. Initial l–l interface 10 mm above the outlet, with a pressure difference of 0.3 bar and oil viscosity of 0.131 Pa·s (cf. Table 2).

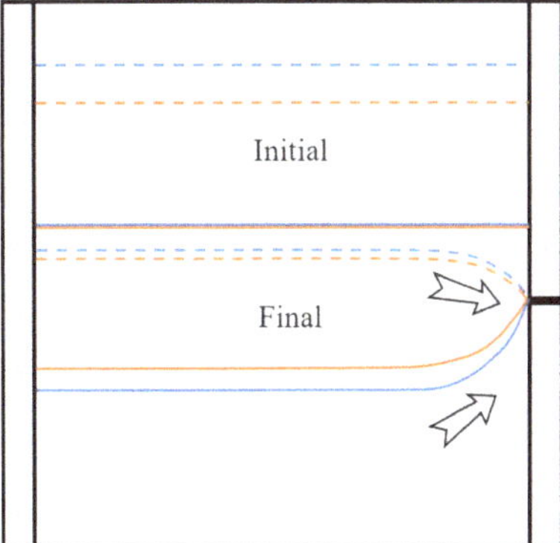

Figure 6. Schematic of initial and final interface states for two initial oil layers enclosed by blue and orange lines. Solid lines: l–l interface, Dashed lines: l–g interface.

3.1.2. Influence of the Initial l–l Interface Level

The initial l–l interface level (above the taphole) is related to the amount of accumulated molten iron in the BF hearth at the commencement of drainage, the dead man porosity, the level of the taphole, and the taphole length. For multi-taphole furnaces, it is also strongly related to the operation of the alternate taphole used [10,12]. The outer level of the taphole is predetermined for every BF, but the angle and the length of the taphole may vary. Even though the former is usually fixed, it is known that some plants used the taphole angle for controlling the drainage. As for the taphole length, it is a controllable factor (under favorable conditions [23]), as it can be adjusted by injecting more or less taphole clay during taphole plugging.

A group of experiments was undertaken to investigate the effect of the initial l–l interface. Figure 7, which illustrates the results, shows how the gas breakthrough time increased with the level of the initial l–l interface at a fixed initial oil layer thickness. This is natural since a higher initial water level delays the moment when oil starts flowing out. The lower panel of the figure shows that both the residual water and oil ratios decreased as the initial l–l interface rose. The reason is that an overall increase in the initial l–l level yields a longer drainage time and therefore lowers residual amounts of the two liquids.

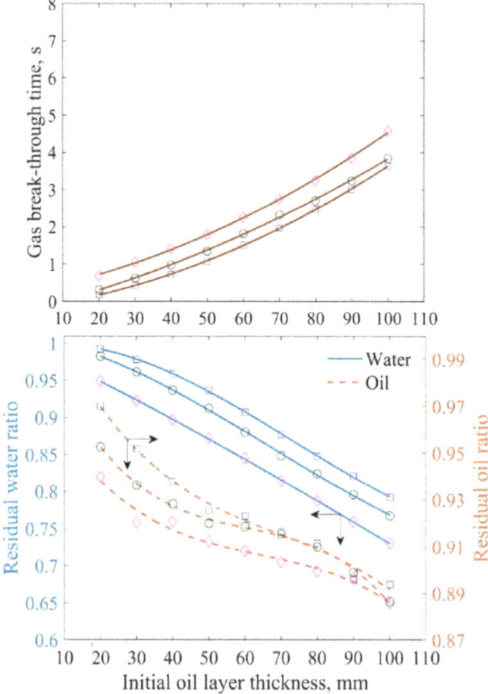

Figure 7. Upper panel: Effect of initial oil layer thickness on gas breakthrough time at different initial l–l interface levels. Lower panel: Effect of initial oil layer thickness on residual ratios of both liquids at different initial l–l interface levels. Pressure difference: 0.3 bar, oil viscosity: 0.131 Pa·s, initial l–l interface levels: 5 mm (marked by squares), 10 mm (marked by circles), and 20 mm (marked by diamonds).

3.1.3. Influence of Pressure Difference

The blast pressure of an operating BF is usually kept relatively stable, which means that the difference between the internal pressure and the atmospheric pressure is constant. However, the pressure difference varies with the blast volume and can also be adjusted by changing the BF top pressure. To clarify the effect of this "driving force", the pressure difference in the experimental system was varied. The results for different oil layer thicknesses are reported in Figure 8, where the upper panel shows that the draining time decreased dramatically with the pressure difference (at fixed initial oil layer thickness and initial l–l level). The obvious reason is that the outflow rates of both liquids increased with the pressure difference. The decrease of the tapping time was more pronounced at a thicker initial oil layer. When the initial oil layer was thin, water was the dominant phase to be drained during the whole drainage period but along with the increase in the initial oil layer thickness, the ratio of water to oil in the outflow decreased, particularly in the later stages of the process. When the initial oil layer was thicker, the tapping time increased significantly due to the overall higher drainage resistance. The lower panel of Figure 8 shows that both residual liquid ratios increase with the increase in pressure difference at a fixed initial amount of water and oil. This was presumably caused by the fact that a higher pressure proportionally increases the draining rate of oil more than that of water, implying an earlier end point of the draining. Thus, a lower pressure difference (i.e., a lower blast pressure) was beneficial for extending the draining time and decreasing the residual liquid ratios. However, in the BF a certain blast pressure is needed for a normal operation of the process to support the burden and to suppress undesired gasification reactions and fluidization.

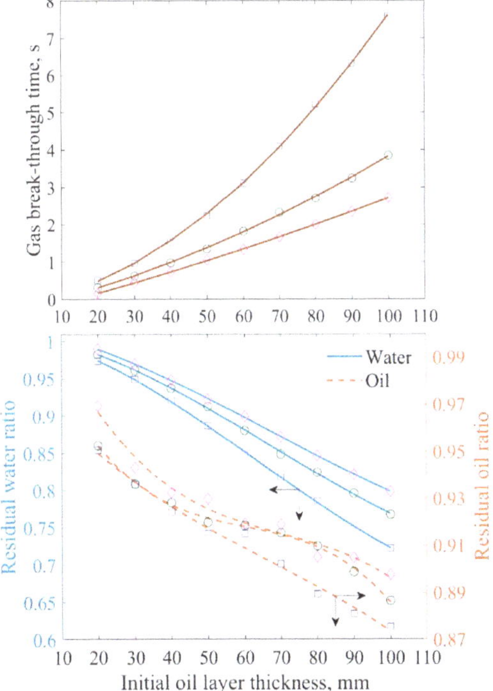

Figure 8. Upper panel: Effect of initial oil layer thickness on gas breakthrough time at different pressure differences. Lower panel: Effect of dimensionless initial oil layer thickness on residual ratios of both liquids at different pressure differences. Initial l–l interface level: 10 mm, oil viscosity: 0.131 Pa·s, pressure differences: 0.2 bar (squares), 0.3 bar (circles), and 0.4 bar (diamonds).

3.1.4. Influence of Oil Viscosity

The slag viscosity mainly depends on temperature and composition, but also on the contents and characteristics of possible solid components (e.g., char). The temperature varies with the thermal state of the process, while the slag composition depends on the type and quantity of raw materials used, and thus it may vary with time. The slag composition is also affected by the partition reactions between iron and slag, which are influenced by the temperature. Figure 9 illustrates the role of oil viscosity on the gas breakthrough time and residual ratios. The tapping time grew with the oil viscosity since the draining rate became lower. With the increase in oil viscosity, the declivity of the l–g interface near the taphole should grow because of an increased pressure drop in the oil phase, but the videos from the experiments showed little change in the bending degree of l–g interface near the taphole, so the effect of draining rate was dominant for this group of experiments. Thus, as the oil viscosity increased, the tapping time increased almost solely due to a decreased tapping rate.

The lower panel of Figure 9 shows that the residual water and oil ratios decreased with an increase in oil viscosity, which is mainly because of the extension of draining time. The results indicate that high slag viscosity may extend draining time, but the high draining resistance and low tapping rate could in practice cause drainage difficulties with an adverse effect on the stability of the hearth operation.

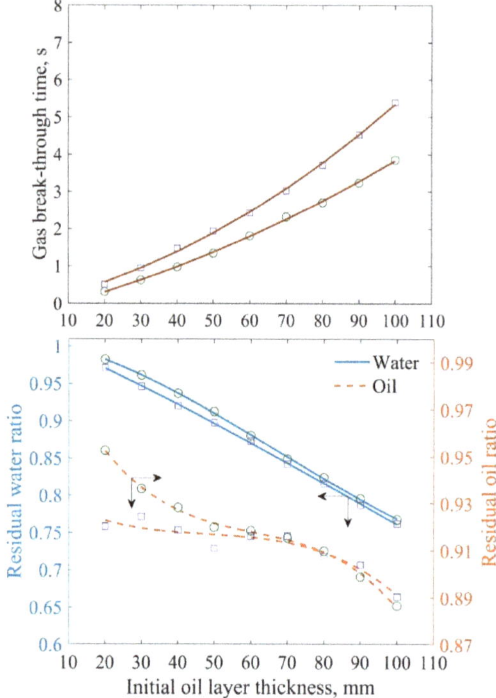

Figure 9. Upper panel: Effect of initial oil layer thickness on gas breakthrough time at different oil viscosities. Lower panel: Effect of initial oil layer thickness on residual ratios of both liquids at different oil viscosities. Initial l–l interface level: 10 mm, pressure difference: 0.3 bar, oil viscosity: 0.131 Pa·s (circles), 0.254 Pa·s (squares).

3.2. Effect of the Factors on the End State

In the draining experiments, three different types of end states occurred, here labelled abnormal drainage, transitional drainage, and normal drainage, respectively. These are illustrated in Figure 10 for three draining experiments using the parameters of experimental group 1, i.e., an initial l–l level of 10 mm, a pressure drop of 0.3 bar, and an oil viscosity of 0.131 Pa·s, but different initial oil layer thicknesses. The horizontal blue line in the figure represents the level of the outlet. In the first type of drainage (top panel, 20 mm initial oil layer), the conditions were such that the overall water level did not even descend to the outlet level but was slightly above it at the point when air broke out. This pattern primarily occurred when the initial oil layer was thin. By contrast, in the final type, i.e., normal drainage (bottom panel, 70 mm initial oil layer), the l–l interface descended clearly below the outlet in the final state and thus bent upwards to the outlet. The cases referred to as transitional represent drainage patterns that fell between these two extremes (middle panel, 40-mm initial oil layer), where the l–l interface was practically horizontal and its vertical level was very close to the outlet level when the drainage ended. It may be noted that this pattern could also occur for a case with a thin initial oil level and an initial l–l interface at the taphole (not studied in the present investigation).

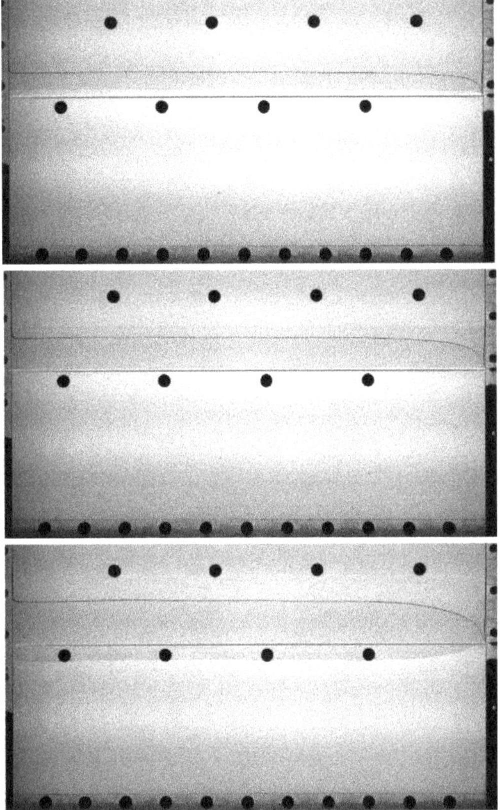

Figure 10. Interface levels at the termination of drainage for abnormal drainage (top panel), transitional drainage (middle panel), and normal drainage (bottom panel) in three experiments with an initial l–l interface level of 10 mm, a pressure drop of 0.3 bar, and oil viscosity of 0.131 Pa·s. The initial oil layer thicknesses are 20 mm, 40 mm, and 70 mm (top to bottom panels).

In order to gain an understanding of the state of the end interfaces, some experiments will be illustrated, showing the influence of the initial oil layer thickness, initial l–l interface level, pressure difference, and oil viscosity. The drainage was studied for different initial oil layer thicknesses. In every group, it was observed that abnormal drainage occurred if the initial oil layer was thin, but with the increase of the oil layer thickness the drainage type evolved gradually through the transition to the normal drainage patterns, as shown in Figure 11. Thus, for each case, there is an initial oil thickness, henceforth called the "critical oil layer thickness", for which the drainage ends with an l–l interface at the taphole. If the initial thickness exceeds the critical value, normal drainage will occur. Thus, for a case with large critical oil layer thickness, abnormal drainage is more likely to occur than for a case with small critical oil layer thickness. To study the effect in more detail, some more experiments were conducted and the end levels of the horizontal part of both interfaces were determined, with results shown in Table 3. The end state is reported as the final level of the horizontal part of the l–l interface, $h_{l\text{-}l,e}$. The conditions for abnormal, transitional, and normal drainage applied were $h_{l\text{-}l,e} > 1$ mm, $1 \geq h_{l\text{-}l,e} \geq -1$ mm, $h_{l\text{-}l,e} < -1$ mm, respectively.

Table 3. Levels of horizontal parts of the interfaces at the end of drainage under different experimental conditions (cf. Table 2). Values that fall in the transition zone have been indicated by bold values, and lack of experimental results by an em dash (—).

Initial Oil Layer Thickness	Experimental Group											
	1		2		3		4		5		6	
mm	$h_{1\text{-}g,e}$ mm	$h_{1\text{-}l,e}$ mm	$h_{1\text{-}g,e}$ mm	$h_{1\text{-}l,e}$ mm	$h_{1\text{-}g,e}$ mm	$h_{1\text{-}l,e}$ mm	$h_{1\text{-}g,e}$ mm	$h_{1\text{-}l,e}$ mm	$h_{1\text{-}g,e}$ mm	$h_{1\text{-}l,e}$ mm	$h_{1\text{-}g,e}$ mm	$h_{1\text{-}l,e}$ mm
20	28.2	+8.0	26.7	+6.6	30.9	+10.4	25.8	+5.2	31.2	+12.0	25.8	+6.2
30	33.1	+3.4	30.9	+1.3	35.7	+5.7	32.8	+2.8	35.6	+6.4	30.6	+1.4
35	35.6	+1.3	32.8	−1.5	—	—	35.6	**+0.2**	—	—	33.1	**−1.0**
40	37.9	−1.0	34.3	−4.8	40.5	+1.7	37.9	−1.7	39.8	+1.1	35.0	−3.4
45	—	—	—	—	43.4	**−0.6**	—	—	42.1	**−2.1**	—	—
50	42.5	−6.0	37.6	−11.1	45.6	−3.3	43.5	−5.7	44.1	−4.3	39.9	−8.2
60	46.6	−12.2	40.4	−18.7	50.5	−7.7	48.1	−11.7	48.0	−10.2	44.7	−13.3
70	50.8	−18.1	42.6	−25.7	55.5	−13.4	51.6	−17.2	51.8	−16.9	49.0	−18.8
80	54.2	−23.5	45.1	−31.9	59.3	−18.6	55.7	−23.7	55.6	−22.5	52.8	−24.5
90	58.4	−29.1	47.5	−39.1	63.6	−24.5	58.8	−28.8	59.7	−28.8	56.9	−30.2
100	60.9	−35.0	50.9	−45.7	68.7	−29.2	63.3	−35.0	62.0	−35.5	60.2	−35.2

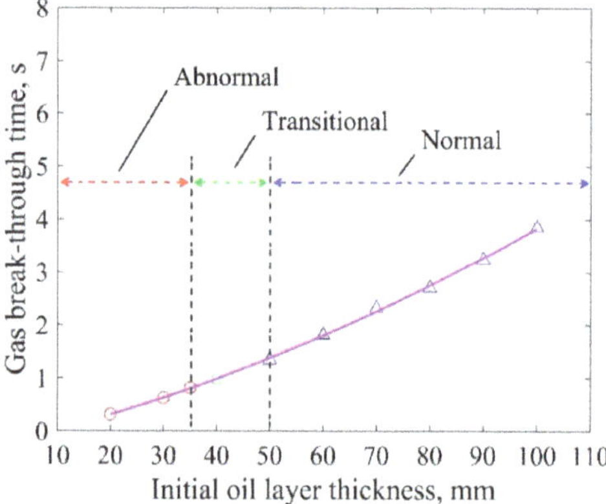

Figure 11. Effect of initial oil layer thickness on the drainage end state. Initial l–l interface level: 10 mm, pressure drop: 0.3 bar, oil viscosity: 0.131 Pa·s.

3.2.1. Effect of the Initial l–l Interface Level

The effect of the initial l–l interface level on the drainage type is presented in dimensional and dimensionless form in Figure 12, demonstrating that the critical oil layer thickness increased and the corresponding flow-out coefficient decreased with the initial l–l interface level. For the purpose of clarity, dashed red lines were drawn through the observations that represent the transitional drainage. A possible explanation for the increase in the critical oil layer thickness is that with the rise of the initial l–l interface, the initial outflow rate of water increases since water occupies the regions above and below the outlet, which makes the oil layer more prone to bend when the oil outflow commences.

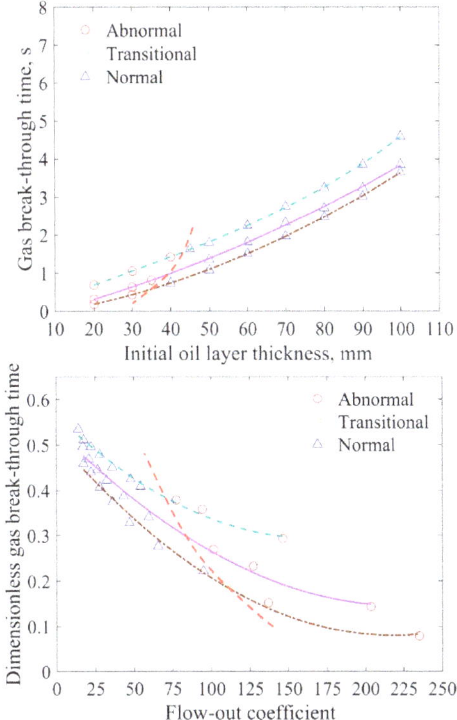

Figure 12. Effect of initial l–l interface level on the drainage end state. Pressure difference: 0.3 bar, oil viscosity: 0.131 Pa·s, initial l–l interface levels: 5 mm (dash-dotted line), 10 mm (solid line), 20 mm (dashed line). Red dashed line shows the states corresponding to transitional drainage.

3.2.2. Effect of the Pressure Difference

The effect of the pressure difference on the end state is illustrated in Figure 13, which shows that the critical oil layer thickness increased with the imposed pressure difference. As the driving force for both water and oil transport increased, the draining rates of both liquids increased, raising the descending speed of the l–l and l–g interfaces. The higher oil draining rate also increased the bending of the l–g interface at the outlet, leading to a decrease in the gas breakthrough time. Consequently, the critical oil layer thickness increased. The lower panel of Figure 13 shows that the flow-out coefficient corresponding to the critical oil layer thickness was rather insensitive to the pressure difference.

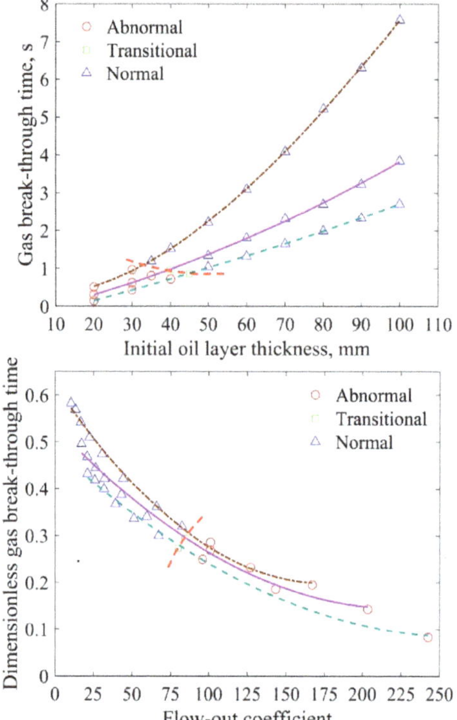

Figure 13. Effect of pressure difference on the drainage end state. Initial l–l interface level: 10 mm, oil viscosity: 0.131 Pa·s, pressure differences: 0.2 bar (dash-dotted line), 0.3 bar (solid line), 0.4 bar (dashed line).

3.2.3. Effect of Oil Viscosity

Finally, Figure 14 illustrates the role of oil viscosity on the drainage type, showing that the critical oil layer thickness decreased and the corresponding flow-out coefficient increased with the oil viscosity. The reason for the decrease of the critical oil layer thickness is that the increased oil viscosity made this phase flow out at a much lower speed, making it less likely that the drainage would end before the water level descended below the taphole.

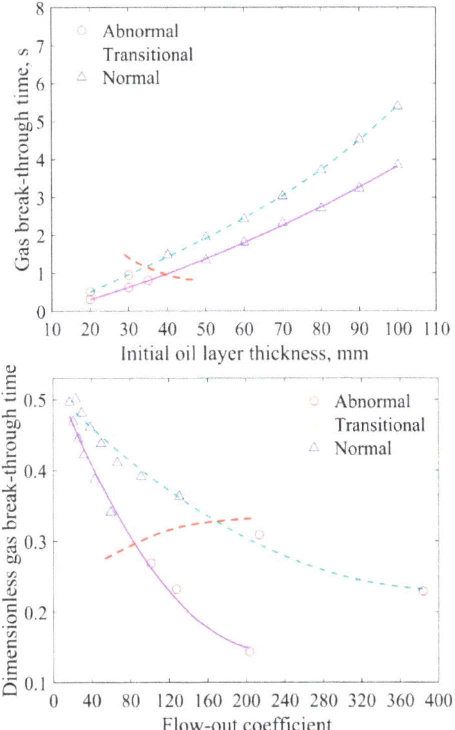

Figure 14. Effect of oil viscosity on the drainage end state. Initial l–l interface level: 10 mm, pressure difference: 0.3 bar, oil viscosity: 0.131 Pa·s (solid line), 0.254 Pa·s (dashed line).

3.3. Comparison with Earlier Findings

To allow for a comparison of the results of the present work with earlier findings by other authors, the residual oil ratios were re-calculated according the definition applied by Tanzil et al. [4,5], i.e., as the ratio of the volume of liquid (slag) remaining above the taphole level at the termination of the drainage to the volume originally above the taphole. Using this definition, the residual oil ratios in the experiments of the present work are depicted in Figure 15. The ratio is seen to increase quite linearly with the logarithm of the flow-out coefficient but levels out at high values. The overall trend is in accordance with that presented by Tanzil and co-workers (e.g., Figure 4 of [4]) even though the flow-out coefficient of the present work is higher, caused by the different experimental conditions (e.g., liquid properties).

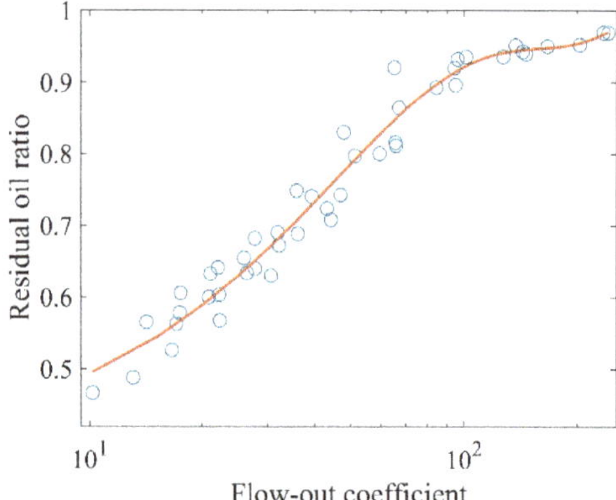

Figure 15. Effect of the flow-out coefficient on the residual oil ratio re-calculated according to the definition in [4,5].

Figure 16 provides a comparison with some other results reported by earlier investigators. The left panels show findings reproduced from He et al. [9] who used a packed bed model in their research, while the right panels are compiled results from the present research for a pressure drop of 0.3 bar and an oil viscosity of 0.131 Pa·s. The different experimental set-up and conditions (fluids, pressure drop) in the two systems make the absolute values of the variables different, so the focus should rather be on trends in the comparison. The upper panels show how the gas break-out time varies with the initial level of the l–g interface. He et al. started the experiments with the l–l interface below (−20 mm), at (0 mm), or above (+20 mm) the outlet, while the experiments of this paper were started above (+5 mm, +10 mm, and +20 mm) the outlet. As observed by He et al., there is a slightly superlinear relation of the gas breakthrough time to the initial l–g level. In contrast to He's results, the present work shows little effect of the initial l–l level. One would, in fact, expect the initial l–l level to have a lowering effect on the drainage time since the drainage is faster if a higher share of the drained liquids is the less viscous one. The differences in drainage time between the experiments with different initial l–l levels are, however, so small that stochastic effects may mask these.

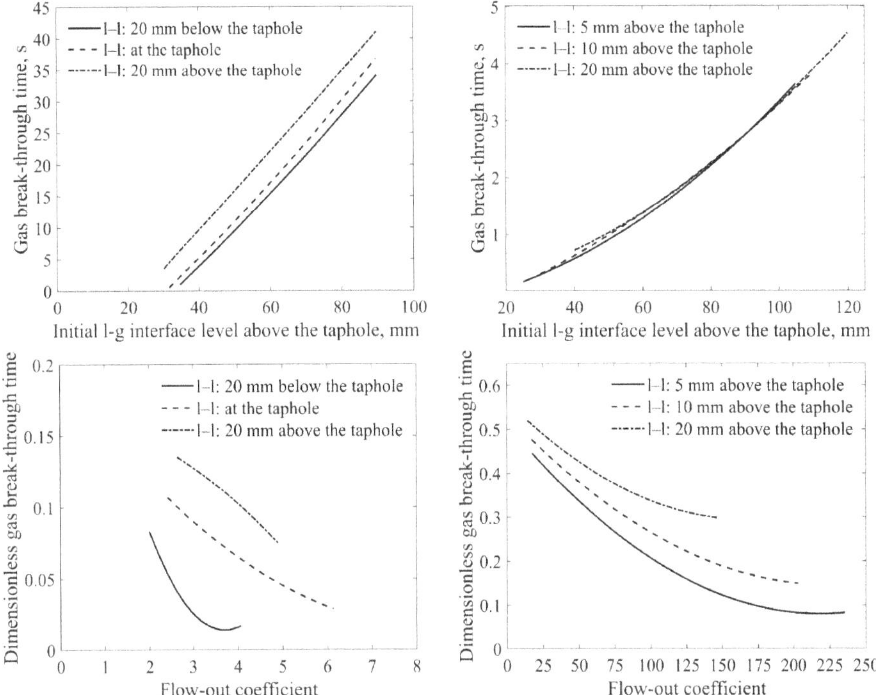

Figure 16. Comparison of some results from He et al. [9] (left panels) with results of the present work (right panels). Top: Effect of initial l–g interface on gas breakthrough time. Bottom: Effect of flow-out coefficient on dimensionless gas breakthrough time.

The lower panels depict how the dimensionless breakthrough time (cf. Equation (8)) depends on the flow-out coefficient. Again, even though the magnitudes of the variables are different, the overall trend exhibited by them can be concluded to coincide well.

4. Conclusions

In order to gain a better understanding of the drainage and evolution of the liquid–gas and liquid–liquid interfaces in the blast furnace hearth, a series of experiments were conducted with a two-dimensional Hele–Shaw model. Oil and water were used as the two liquids emulating slag and iron in the real system. Special attention was focused on the effect of the conditions on the gas breakthrough time, residual ratios of both liquids, and the end state of the liquid levels, categorized into three groups (normal, transitional, or abnormal drainage). Based on the experimental results, the following conclusions can be drawn:

(1) An increase in the initial oil layer thickness (i.e., more accumulated slag in the hearth) extends the draining time and decreases the residual ratios of both liquids. In addition, the likelihood of abnormal drainage also decreases.

(2) With a higher initial l–l interface level, the draining time increases and the residual liquid ratios decrease. The likelihood of abnormal drainage increases.

(3) With an increased pressure difference ("driving force") in the system, the residual liquids ratios and the likelihood of abnormal drainage increase, while the gas breakthrough time decreases.

(4) An increase of oil viscosity increases the tapping time and reduces the likelihood of abnormal drainage and residual ratios of both liquids.

The findings can be used to provide a better interpretation of the drainage patterns observed in the blast furnace, and the results may further be applied to calibrate and verify computational fluid dynamics models of the blast furnace hearth.

Author Contributions: Conceptualization, W.L. and L.S.; Data curation, W.L. and L.S.; Formal analysis, W.L. and H.S.; Funding acquisition, H.S.; Investigation, W.L. and L.S.; Methodology, W.L. and L.S.; Project administration, H.S.; Resources, H.S.; Software, W.L.; Supervision, H.S. and L.S.; Writing—original draft, W.L.; Writing—review and editing, H.S. All authors have read and agreed to the published version of the manuscript

Funding: This research was funded by National Natural Science Foundation of China (Grant 51604068) and Åbo Akademi Foundation.

Acknowledgments: This work was carried out with support from the Åbo Akademi Foundation, and this funding is gratefully acknowledged. The second author (Lei Shao) wishes to thank the National Natural Science Foundation of China (Grant 51604068) for providing financial support for this work.

Conflicts of Interest: The authors declare no conflict of interest.

References

1. Geerdes, M.; Toxopeus, H.; vd Vliet, C. *Modern Blast Furnace Ironmaking: An Introduction*, 2nd ed.; IOS Press BV: Amsterdam, The Netherlands, 2009.
2. Brännbacka, J. Model Analysis of Dead-Man Floating State and Liquid Levels in the Blast Furnace Hearth. Ph.D. Thesis, Åbo Akademi University, Turku, Finland, 2004.
3. Pinczewski, W.V.; Tanzil, W.B.U.; Hoschke, M.I.; Burgess, J.M. Simulation of the drainage of two liquids from a blast furnace hearth. *Tetsu-to-Hagane* **1982**, *68*, S111.
4. Tanzil, W.B.U.; Zulli, P.; Burgess, J.M.; Pinczewski, W.V. Experimental model study of the physical mechanisms governing blast furnace hearth drainage. *Trans. Iron Steel Inst. Jpn.* **1984**, *24*, 197–205. [CrossRef]
5. Tanzil, W.B.U. Blast Furnace Hearth Drainage. Ph.D. Thesis, University of New South Wales, Sydney, Australia, 1985.
6. Zulli, P. Blast Furnace Hearth Drainage with and without a Coke-Free Layer. Ph.D. Thesis, University of New South Wales, Sydney, Australia, 1991.
7. Shao, L.; Saxén, H. Simulation study of blast furnace drainage using a two-fluid model. *Ind. Eng. Chem. Res.* **2013**, *52*, 5479–5488. [CrossRef]
8. He, Q.; Zulli, P.; Evans, G.M.; Tanzil, F. Free surface instability and gas entrainment during blast furnace drainage. *Dev. Chem. Eng. Miner. Process.* **2006**, *14*, 249–258. [CrossRef]
9. He, Q.; Evans, G.; Zulli, P.; Tanzil, F. Cold model study of blast gas discharge from the taphole during the blast furnace hearth drainage. *ISIJ Int.* **2012**, *52*, 774–778. [CrossRef]
10. Iida, M.; Ogura, K.; Hakone, T. Analysis of drainage rate variation of molten iron and slag from blast furnace during tapping. *ISIJ Int.* **2008**, *48*, 412–419. [CrossRef]
11. Nouchi, T.; Yasui, M.; Takeda, K. Effects of particle free space on hearth drainage efficiency. *ISIJ Int.* **2003**, *43*, 175–180. [CrossRef]
12. Roche, M.; Helle, M.; van der Stel, J.; Louwerse, G.; Shao, L.; Saxén, H. Off-line model of blast furnace liquid levels. *ISIJ Int.* **2018**, *58*, 2236–2245. [CrossRef]
13. Fukutake, T.; Okabe, K. Influences of slag tapping conditions on the amount of residual slag in the blast furnace hearth. *Trans. Iron Steel Inst. Jpn.* **1976**, *16*, 317–323. [CrossRef]
14. Brännbacka, J.; Torrkulla, J.; Saxén, H. Simple simulation model of blast furnace hearth. *Ironmak. Steelmak.* **2005**, *32*, 479–486. [CrossRef]
15. Brännbacka, J.; Saxén, H. Modeling the liquid levels in the blast furnace hearth. *ISIJ Int.* **2001**, *41*, 1131–1138. [CrossRef]
16. Roche, M.; Helle, M.; van der Stel, J.; Louwerse, G.; Shao, L.; Saxén, H. On-line estimation of liquid levels in the blast furnace hearth. *Steel Res. Int.* **2019**, *90*, 1800420. [CrossRef]
17. Nishioka, K.; Maeda, T.; Shimizu, M. Effect of various in-furnace conditions on blast furnace hearth drainage. *ISIJ Int.* **2005**, *45*, 1496–1505. [CrossRef]
18. Zhou, Z.; Zhu, H.; Yu, A.; Zulli, P. Numerical investigation of the transient multiphase flow in an ironmaking blast furnace. *ISIJ Int.* **2010**, *50*, 515–523. [CrossRef]

19. Vångö, M.; Pirker, S.; Lightenegger, T. Unresolved CFD–DEM modeling of multiphase flow in densely packed particle beds. *Appl. Math. Model.* **2018**, *56*, 501–516. [CrossRef]
20. Bambauer, F.; Wirtz, S.; Sherer, V.; Bartusch, H. Transient DEM-CFD simulation of solid and fluid flow in a three dimensional blast furnace model. *Powder Technol.* **2018**, *334*, 53–64. [CrossRef]
21. Bear, J. *Dynamics of Fluids in Porous Media*, 1st ed.; American Elsevier Company: New York, NY, USA, 1972; p. 133.
22. Liu, W.; Mondal, D.N.; Shao, L.; Saxén, H. An image analysis-based method for automatic data extraction from pilot draining experiments. *Ironmak. Steelmak* **2020**. submitted manuscript.
23. Tsuchiya, N.; Fukutake, T.; Yamauchi, Y.; Matsumoto, T. In-furnace conditions as prerequisites for proper use and design of Mud to control blast furnace taphole length. *ISIJ Int.* **1998**, *38*, 116–125. [CrossRef]

Article

Characterisation of a Real-World Søderberg Electrode

Ralph Ivor Glastonbury [1], Johan Paul Beukes [1,*], Pieter Gideon van Zyl [1], Merete Tangstad [2], Eli Ringdalen [3], Douglas Dall [4], Joalet Dalene Steenkamp [5,6] and Masana Mushwana [5]

1   Chemical Resource Beneficiation, North-West University, Potchefstroom 2520, South Africa; 23032898@student.g.nwu.ac.za (R.I.G.); Pieter.VanZyl@nwu.ac.za (P.G.v.Z.)
2   Department of Materials Science and Engineering, Norwegian University of Science and Technology (NTNU), 7465 Trondheim, Norway; merete.tangstad@ntnu.no
3   Metal Production and Processing, Department of Materials and Chemistry, Sintef, 7465 Trondheim, Norway; Eli.Ringdalen@sintef.no
4   Glencore Alloys, Rustenburg 0300, South Africa; Douglas.Dall@glencore.co.za
5   Pyrometallurgy Division, Mintek, Randburg 2125, South Africa; JoaletS@mintek.co.za (J.D.S.); MasanaM@mintek.co.za (M.M.)
6   Faculty of Engineering & the Built Environment, School of Chemical & Metallurgical Engineering, University of the Witwatersrand, Johannesburg 2000, South Africa
*   Correspondence: paul.beukes@nwu.ac.za; Tel.: +27-18-299-2337

Abstract: Very little research on Søderberg electrodes has been published in the journal peer reviewed public domain. The main aim of this work is to characterise a Søderberg electrode that was cut off approximately 0.5 m below the contacts shoes of a submerged arc furnace. Additionally, the characterisation data can be used to verify if Søderberg electrode models accurately predict important electrode characteristics. The operational history (slipping, current, and paste levels) proved that the case study electrode was a representative specimen. The characterisation results indicated no significant electrical resistivity, degree of graphitisation (DOG), and bulk density changes from 0.7 to 2.7 m on the non-delta side (outward facing), while these characteristics changed relatively significantly on the delta side (inward facing) of the electrode. The area where the submerged arc would mostly like jump off the electrode had the lowest resistivity, as well as highest DOG and bulk density. No significant difference in porosity as a function of length below the contact shoes were observed; however, slight increases occurred near the perimeters. It was postulated that oxidation of carbon resulted in increased pore volumes near the electrode perimeter. No significant difference in compressive breaking strength was observed over the electrode area investigated.

Keywords: Søderberg electrodes; submerged arc furnace (SAF); ferro-alloy production; ferrochrome; electrical resistivity; degree of graphitisation; bulk density; porosity; compressive breaking strength

Citation: Glastonbury, R.I.; Beukes, J.P.; van Zyl, P.G.; Tangstad, M.; Ringdalen, E.; Dall, D.; Steenkamp, J.D.; Mushwana, M. Characterisation of a Real-World Søderberg Electrode. *Metals* 2020, *11*, 5. https://dx.doi.org/10.3390/met11010005

Received: 31 October 2020
Accepted: 18 December 2020
Published: 22 December 2020

Publisher's Note: MDPI stays neutral with regard to jurisdictional claims in published maps and institutional affiliations.

1. Introduction

Since its development in the early 1900s [1], Søderberg electrodes have been used extensively in submerged arc furnaces (SAFs) to conduct electrical current from a transformer(s) to the smelting zone [2]. Various authors have presented representations of a Søderberg electrode column and/or described the processes associated with it, e.g., [2–10]. Such a column typically consists of a cylindrical steel casing that extends from a building level well above the furnace, down into the smelting zone. The casing serves as the mould for the electrode paste, which contains a binder (e.g., coal tar pitch) mixed with a solid aggregate (e.g., calcined anthracite or coke), to form the electrode during operation. The casing also has longitudinal fins, which add dimensional stability to the casing and the unbaked electrode. Resistive heating of current passing through the casing, fins, and paste, as well as heat conducted upward from the SAF in the electrode column melt the electrode paste at temperatures above its softening point to fill the casing. Electrode paste softening temperatures have been reported to be in the range of 65 to 134 °C [11]. Solid electrode paste is added at regular intervals via the top of the electrode casing to maintain a steady level of solid (un-melted) and liquid paste (that have filled the casing). The casing is slipped (lowered in relation to clamping devices that hold the electrode) relatively small distances on a shift basis (usually 8 h long), and longer slips are taken if an electrode break occurs, or if the electrode is shorter than ideal. All the afore-mentioned result in the melted paste being exposed to increasing temperatures as it approaches the so-called contact shoes. Contact shoes are large copper (or copper alloy) components that are pressed against the casing to conduct electrical current into the electrode. Before the liquid electrode paste reaches the bottom of the contact shoes, it should ideally be transformed into solid carbonaceous material. The temperature at which the paste transforms from a liquid to a solid is termed the baking isotherm. This transformation temperature has been reported to be between 450 and 475 °C [11]. As the carbonaceous electrode descends further into the SAF temperatures increase, which results in further baking.

Good Søderberg electrode management, together with other aspects such as maintaining a suitable metallurgical balance (to obtain product within specifications and the correct slag chemistry), high electrical power input (assuming fixed specific energy consumption, production volume is directly related to power input), and feeding the furnace appropriately (without feed it cannot produce and asymmetrical feed could lead to damage) form the basis for safe (occupational health and safety of workers), environmentally responsible (limiting emissions), and profitable smelting operations. However, good Søderberg electrode management implies that a multitude of actions and tracking of parameters need to take place on a shift and daily basis.

Very harsh conditions in the SAFs lead to severe mechanical and thermal stresses, which result in unavoidable wear of the electrode tips. Furthermore, thermal stresses associated with shutdowns and operational instability may cause anything from inconsequential hard electrode breaks to disastrous green breaks [7,8]. Broken electrodes can negatively impact process productivity (e.g., lower electrical power input, damage to equipment, or even prolonged shutdowns to repair damage), personnel occupational health and safety, as well as environmental emissions (e.g., process instability often result in higher than normal atmospheric emissions).

Notwithstanding the importance of Søderberg electrode design and management to ensure responsible, safe, and profitable smelting operations, very little research focussing specifically on Søderberg electrode related aspects have been published in the journal peer reviewed public domain. For instance, as far as the authors could assess (using the "Scopus" scientific search engines with the key words "Soderberg electrode"), only 5 such papers have been published in the last decade (i.e., 2011 to 2020) [4,11–14]. Several conference proceeding papers have also been published during this time, e.g., [2,3,5,15]. These conference proceeding papers also gave valuable topic specific insights. However, many such contributions were made principally to demonstrate the capability of commercial companies to serve their clients instead of divulging all relevant information as is common in academic journal papers.

The main aim of this work was to characterise a largely intact Søderberg electrode, removed from a ferrochrome SAF during a shutdown to rebuild many of the furnace components. The operational history for a month prior to the shutdown, procedures followed to sample the electrode and experimentally determined characteristics (e.g., electrical resistivity, degree of graphitisation, bulk density, porosity, and compressive breaking strength) are presented. As far as the authors could assess, such a characterisation of a real-world Søderberg electrode have not yet been presented in the peer reviewed public domain. Such data will contribute towards a better understanding of electrode management by operational personnel and assist researchers in identifying unresolved questions. Additionally, the characterisation data can be used to verify such characteristics in Søderberg electrode models, which have up to now mainly been restricted to temperature, current densities, and thermal stress profiles, e.g., [2,3,5,6].

2. Materials and Methods

2.1. Søderberg Electrode Sampling

On 31 July 2013, an operational ferrochrome SAF with 1.4 m casing outer diameter Søderberg electrodes was switched off, after following the appropriate preparations for a furnace rebuild. The furnace was allowed to cool down for 5 days. Thereafter, the number 2 electrode (Figure 1a indicating the numbering from a top view), was cut off with a diamond wire saw approximately 0.5 m below the contact shoes, as indicated in Figure 1b. The electrode stump was removed from the furnace with appropriate rigging equipment. Then, 100 mm outer diameter core drill samples were collected, with the 100 mm sample centred at 0.70, 1.15, 1.64, 2.16, and 2.70 m below the contact shoes. This core drilling was done through the entire diameter of the electrode, from the outside towards the delta side of the electrode. Here the term "delta" refers to the side of the electrode facing the centre of the SAF (see Figure 1a). An example of such a core drilled sample is presented in Figure 1c. The 100 mm diameter core samples were then cut into 10 mm disks (examples indicated in Figure 1d) with a commercial cut off grinder that was modified for the purpose. Precise lathe turned; tight fitting bushings were installed in the pivot point of this cut-off grinder to ensure accurate cutting. Additionally, it was permanently fitted on a table with specially designed cradles to hold the 100 mm tube-like core drilled electrode sample perpendicular to the blade of the cut-off grinder. Cylindrical sample pellets with outer diameters of 8 mm were then drilled out from these 10 mm thick disks with an appropriate diamond coated core drill fitted in a bench drill, as indicated in Figure 1e,f. The condition of a specific 100 mm thick disk determined how many such 8 mm × 10 mm cylindrical sample pellets could be obtained from it, with ≥10 being the norm. These 8 mm × 10 mm cylindrical sample pellets were used in all electrode characterisation experiments, unless stated otherwise. Since many of the characterisation techniques involved destructive analysis, the investigated characteristics were not performed on material sampled from the same positions within the case study electrode. Therefore, it was not possible to directly correlate (via correlation graphs) characteristics with one another.

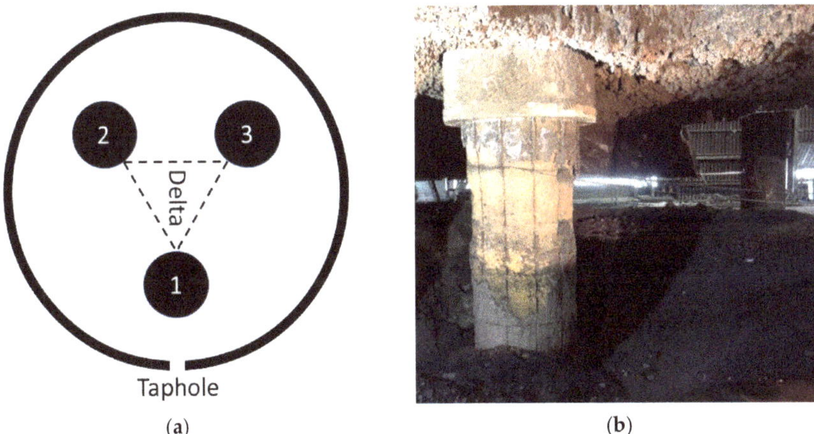

(a)   (b)

Figure 1. *Cont.*

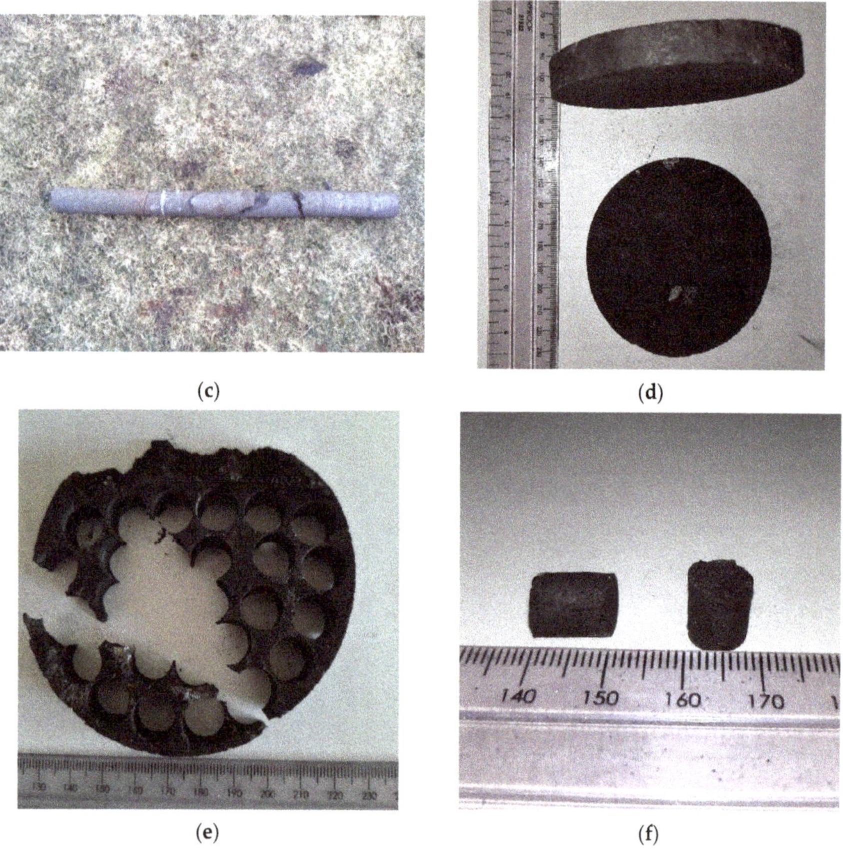

(c)

(d)

(e)

(f)

Figure 1. (**a**) Diagram from a top view indicating the electrode numbering of the SAF from which the electrode was removed; (**b**) photo indicating how the electrode was cut off, approximately 0.5 m below the contact shoes with a diamond wire saw; (**c**) an example 100 mm outer diameter core drill sample, collected by core drilling through the entire diameter of the electrode, from the outside towards the delta side of the electrode; (**d**) examples of 10 mm thick disks that were cut from the 100 mm diameter core samples; (**e**) example 10 mm thick disk, after 8 mm outer diameter cylindrical sample pellets were drilled out; (**f**) example 8 mm × 10 mm cylindrical sample pellets that were used in most experiments.

2.2. Electrical Resistivity Measurements

The electrical resistivities of cylindrical sample pellets were measured at room temperature using a GW Instek resistance precision meter (Model LCR-6020 Series). The meter can be used to measure inductance (L), capacitance (C), and resistance (R) of samples over a wide frequency (10 Hz–20 kHz), resistance (0.99–99.9 MΩ), and voltage range (10.00 mV–2.00 V). As indicated in Figure 2a, a two-probe device was manufactured (by the North-West University Instrument Makers) similar to the design presented by Heikkilä et al. [16] and connected to the LCR meter. This two-probe device had a bottom and a top probe, which were both turned from solid copper. The stainless steel tightening wheel mechanism (Figure 2a) was used to gently tighten a cylindrical sample pellet in-between these copper probes prior to a measurement. A torque wrench was then used to fasten the hex/Allen key located on top of the stainless steel tightening wheel mechanism further. This ensured that each sample was set under the same torque pressure (i.e., 0.5 Nm). The crocodile clamps of the LCR meter were clamped into fit-for-purpose holes drilled in the copper probes. These holes ensured consistent placement of the clamps on the probes, as well as consistent contact area between the clamps and the probes. The frame and connectors that attached the probes to the frame were made of polyvinyl chloride (PVC) plastic, to isolate the frame from the sample. As illustrated in Figure 2b, the face of the bottom copper probe on which the cylindrical sample pellet sat had a small rim on the outside, while the corresponding face of the top copper probe was flat. These shapes were used to ensure that the cylindrical sample pellets were held in-between the two copper probes in a repeatable manner. Both copper probes were also set on a 360° articulating spring mechanism. This kept the ends of the cylindrical sample pellet perpendicular to the probe faces, even if the cylindrical sample pellet ends were not perfectly perpendicular to the cylindrical sample pellet sides.

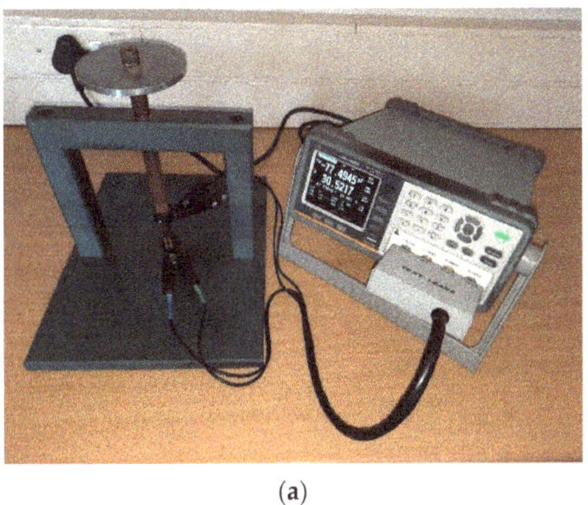

(a)

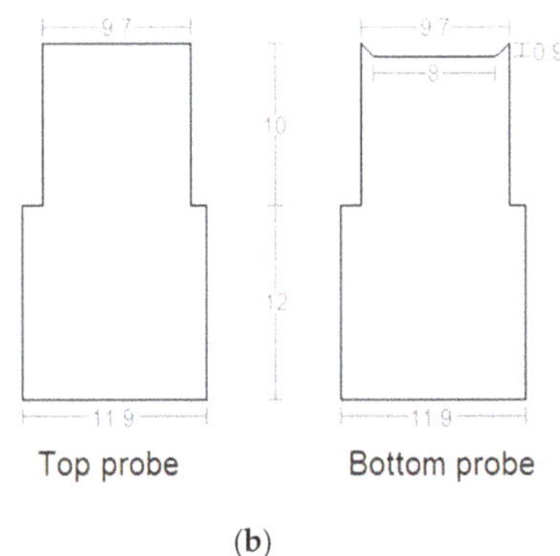

(b)

Figure 2. (a) Photo indicating the GW Instek resistance precision meter (Model LCR-6020 Series) and two-probe device, similar to the design presented by Heikkilä et al. [16], used to measure electrical resistivity of the cylindrical sample pellets; (b) illustration indicating the shapes of the copper probes, in-between which the cylindrical sample pellets were held.

To ensure precision and accuracy, the LCR meter was switched on for 30 min and a daily calibration was conducted before measurements commenced. Cylindrical copper, aluminium, and stainless steel samples, which had dimensions similar to the cylindrical sample pellets, were measured during calibration. The calibration values were stored by the LCR meter to correct for errors caused by impedance of the experimental setup. The resistance values of the measured cylindrical sample pellets were calculated from Ohm's law:

$$R = V/I, \tag{1}$$

where R is the resistance (Ω), V is the voltage across the sample (V), and I is the measured current (A). Assuming uniform current distribution, the resistivity was calculated by Equation (2):

$$\rho = (V.A)/(I.H), \tag{2}$$

where ρ is the material resistivity ($\Omega \cdot m$), V is the voltage across the sample, I is the measured current (A), A is the cross-sectional area of the cylindrical sample (m^2), and H is the height of the sample (m).

2.3. Degree of Graphitisation Determinations

The degree of graphitisation (DOG) was determined with the so-called interlayer spacing or d_{002} method, based on X-Ray Diffraction (XRD) analysis [17]. This method is one of at least three recognised methods to determine DOG [18]. The powder generated during drilling of the 8 mm × 10 mm cylindrical sample pellets (as explained in Section 2.1) were used for this purpose. A small amount of silicon metal powder, used as a reference material, was added to the powdered Søderberg electrode material and mixed thoroughly, prior to analysis. The prepared samples were analysed on a Bruker D8 A25 DaVinci X-ray Diffractometer and LynxEye™ SuperSpeed Detector with xyz-stage mounted and using Cu-K radiation. The DOG was calculated with Equation (3) [17]:

$$DOG = (3.440 - d_{002})/(3.440 - 3.354) = (3.440 - d_{002})/0.086, \qquad (3)$$

where DOG is the degree of graphitisation (unitless, with 1 indicating fully graphitised and 0 fully un-graphitised), 3.440 is the interlayer spacing of fully un-graphitised carbon (nm), 3.345 is the interlayer distance for pure graphite (nm), and d_{002} is the interlayer spacing derived from XRD analysis (nm).

2.4. Bulk Density Measurements

Bulk densities were measured using the ASTM C20-00 method [19]. Cylindrical sample pellets with 18 mm × 10 mm dimensions, obtained in a similar manner as the 8 mm × 10 mm cylindrical sample pellets (described in Section 2.1), were used in these determinations. The samples were first air dried at 105 °C for 3 h in an oven. After cooling in a desiccator, the samples were weighted (D). They were then saturated by boiling them in distilled water for 2 h. Thereafter, the samples were weighted in water to get the saturated weight (S). The bulk density was calculated with the following formula:

$$\rho = [D/(D - S)(\rho o - \rho L)] + \rho L, \qquad (4)$$

where ρ is the bulk density of the sample (g·cm^{-3}), D the sample weight after drying, S the saturated sample weight, ρo the density of water (taken as 0.9982 g·cm^{-3}), and ρL the density of air (taken as 0.0012 g·cm^{-3}).

Bulk densities were also measured with an AccyPyc II 1340 gas pycnometer. Søderberg electrode pieces remaining after the 8 mm × 10 mm cylindrical sample pellets were drilled out of the 10 mm thick disk (Figure 1e) were used for this purpose. Each sample was first weighed and then placed in a 35 cm^3 sample holder in the instrument chamber. The chamber was purged 5 times with 99.99% helium. A known quantity of gas at a known pressure was then expanded into the chamber and the new pressure measured. The difference between the initial and final pressures and the known volume of the chamber is then used (with the gas law) to calculate the volume of the sample. And with the sample mass, the sample bulk density could be calculated.

2.5. Porosity Determinations

Computerized tomography (CT) scanning was used to generate sample image slices, which could be processed to determine porosity. A Nikon XT H 225 ST with a 225 kV reflection target source and a Perkin Elmer 1620 detector CT instrument was used. Similar to the bulk density determinations (described in Section 2.4), Søderberg electrode pieces remaining after the 8 mm × 10 mm cylindrical sample pellets were drilled out of the 10 mm thick disk (Figure 1e) were used. The samples were placed in glass poly tops stacked 6 high and wrapped in cellophane to keep them aligned and easy to handle. Each poly top was scanned individually, with 2000 sample image slices being generated per poly top. From these 2000 image slices per sample, 10 images which were evenly spaced (every 160 images between images 200 and 1800) were used for the actual porosity determinations. An example of a CT image slice is presented in Figure 3a. ImageJ (https://imagej.nih.gov/ij/, accessed 20/10/2020) was used to analyse the CT image slices. This was done by adjusting the threshold value so that the pores and background was red in colour (Figure 3b) in all images. In each image, the boundaries of the solid particle(s) visible in the CT image slice were then manually marked (indicated by the yellow line and white markers in Figure 3b). Thereafter, the measurement tool of ImageJ was used to determine the surface area of red pixels within the selected area(s), as a percentage of all the pixels within the selected area(s). This corresponded to the percentage pores in the particular CT image slice.

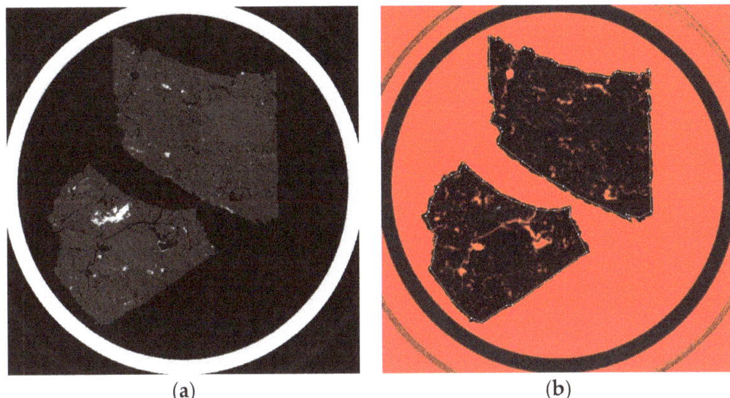

Figure 3. (**a**) An example of a CT image slice; (**b**) ImageJ (https://imagej.nih.gov/ij/, accessed 20/10/2020) was used to analyse the CT image slices. The threshold value was adjusted so that the pores and background was red in colour. The boundaries of the solid particle(s) visible in the CT image slice were manually marked (indicated by the yellow line and white markers).

2.6. Compressive Breaking Strength Measurements

The compressive breaking strengths of 8 mm × 10 mm cylindrical sample pellets were determined using an Ametek Lloyd Instruments LRXplus 5 kN strength tester. NEXYGEN-Plus material test and data analysis software was used to control the instrument and capture the generated data. The speed of the compression plates moving towards one another was kept constant at 10 mm·min^{-1} during all compressive strength tests. The 8 mm × 10 mm cylindrical sample pellets were placed upright in the breaking strength machine (flat sample faces on compression plates), and the maximum load to induce breakage of each sample pellet was recorded.

2.7. Statistical Handling of Data

For electrical resistivity, between 10 and 30 of the 8 mm × 10 mm cylindrical sample pellets from a single 10 mm thick sample disk was measured, and each such pellet was measured 15 times. DOG determinations were only done once on each sample considered. In the ASTM C20-00 density experiments, 2 of the 18 mm × 10 mm sample pellets from a single 10 mm thick sample disk were considered, and each such pellet was used only once. In the gas pycnometry analyses, cuts off of sample pellets were used, and each analysis was repeated 5 times. Porosities based on CT scanning and image processing was conducted as indicated in Section 2.6. For compressing breaking strength, between 10 and 30 of the 8 mm × 10 mm cylindrical sample pellets were tested from each considered 10 mm thick sample disk.

Where multiple repetitions were done, outliers were identified with a commonly applied method (e.g., https://www.mathworks.com/help/stats/box-plots.html, accessed 30/10/2020). This was done by multiplying the interquartile range (IQR) with 1.5, and datapoints that were above the 75th quartile plus the IQR and below the 25th percentile minus the IQR were disregarded in further calculations. Where applicable, standard deviations were presented with average values to indicate data repeatability.

3. Results and Discussion

3.1. Operational History

The results presented in this paper will only be relevant to Søderberg electrode use in SAF operations, if the case study electrode represented normal/stable operation. To explore this, total daily slipping taken on the electrode, daily average current passing through it, as well as daily average solid and liquid paste levels in the month prior to the furnace shutdown, are presented in Figure 4. From this data, it is evident that the slipping was relatively consistent (varied between 14 to 32 mm per day). No long slips (usually defined as >100 mm) were taken on this electrode, which indicates that the operational personnel considered its length as normal and that no breaks occurred during this 1 month period. The daily average current passing through the electrode varied between approximately 63 and 76 kA, and had an average of 72.2 kA in the period between 1 and 28 July 2013. This again indicates that no major operational issues occurred. In the last three days prior to the shutdown (29 to 31 July 2013) the energy input was purposefully reduced, which resulted in the average current passing through the electrode on the last operational day being 58.0 kA. Such a gradual reduction in power input contributes significantly to prevent thermal shock on the electrode [5]. Average daily solid (varying between 6.7 and 9.3 m) and liquid (varying 2.9 and 4.1 m) paste levels were normal, again with no indication that paste consumption increased or decreased significantly during the last month of operation. Considering all the afore-mentioned, in conjunction with visual inspection of the electrode and its length before being cut off (it was approximately 3 m long), it can be concluded that this case study electrode was representative of stable SAF operation.

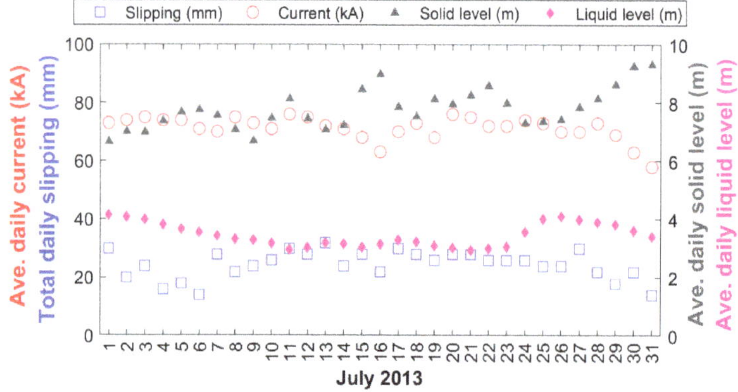

Figure 4. Total daily slipping taken on the case study electrode, daily average current passing through it, as well as daily average solid and liquid paste levels in the month prior to the furnace shutdown.

3.2. Electrical Resisitivity

The main function of Søderberg electrodes is to conduct electrical current into the smelting zone of the furnace. It is therefore fitting that electrical resistivity is the first characteristic that is considered. Figure 5a presents the average measured resistivity and linear fitting of this data, as a function of length below the contact shoes and distance from the delta side of the electrode. It should be noted that no experimental data could be collected on, or very close to 0 and 1.4 m from the electrode delta side, since the 100 mm core drilled samples (Figure 1c) were typically ≤1.3 m long. This smaller electrode diameter than the original outer casing diameter of 1.4 m (introduced in Section 2.1), is due to several factors. Firstly, the mild steel casing usually burns off and/or melts within the first approximately 0.5 m below the contact shoes, which is also shown in Figure 1b for the specific electrode considered here. The casings used on the case study electrode was 3 mm thick. Therefore, 6 mm (3 mm on each side) can be attributed to the casing loss. Secondly, the electrode paste will shrink at temperatures above the baking isotherm [5]. This occurs since the carbonaceous material structure becomes more organised and closer to that of graphite, due to the increased temperature the electrode experiences as it descends. Thirdly, chemical corrosion, i.e., oxidation of the electrode carbon above the bed material, can also contribute, especially since the furnace considered here was an open-SAF. In open-SAFs, atmospheric air containing oxygen can relatively freely enter (determined by off-gas flow rate) and results in a partially oxidising environment above the bed material [20]. Lastly, physical erosion, due to frictional and mechanical forces exerted on the electrode, can also erode its outer perimeter. Due to all the afore-mentioned reasons, all result figures wherein the case study electrode diameter is specified will indicate it as 1.3, not 1.4 m.

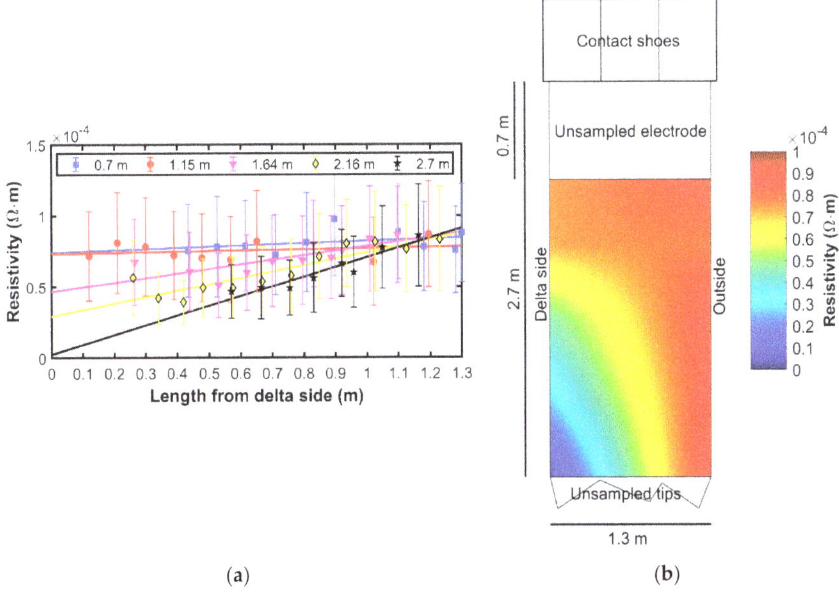

Figure 5. (a) Average measured electrical resistivity (markers) and linear fitting of this data (lines), as a function of length below the contact shoes and distance from the delta side of the electrode; (b) 3D rendering of the fitted line data associated with resistivities at 10 mm intervals across the electrode at each of the lengths below the contacts shoes indicated in Figure 5a.

The standard deviations of the resistivity measurements presented in Figure 5a were relatively large. This was mostly due to the actual resistivities being so low (i.e., 10^{-5} to 10^{-6} $\Omega \cdot$m). Therefore, to ensure that the average values could be used to identify trends, very large numbers of measurements (i.e., N = 150 to 450) were conducted. Considering the average values, Figure 5a clearly indicates that the electrical resistivity on the outside (non-delta side) of the electrode between 0.7 and 2.7 m below the contact shoes, were essentially consistent. In contrast, the resistivity decreased (implying increased conductivity) significantly from 0.7 to 2.7 m below the contact shoes on the delta side of the electrode. To visualise the electrical resistivity results better, the average data associated with the fitted lines presented in Figure 5a at 10 mm intervals across the diameter of the electrode at each of the investigated lengths below the contacts shoes were spatial interpolated across the entire investigated area (0.7 to 2.7 m below the contacts shoes), as indicated in Figure 5b. This was done using the "grid data" function in Matlab, with triangulation-based linear interpolation. In effect, this representation gives a 3-dimensional (3D) cross section perspective of the investigated electrode, excluding the unsampled electrode area directly below the contact shoes and the uneven electrode tips, with colour indicating the 3rd dimension. This 3D perspective indicates that resistivity was at its lowest at the bottom end on the delta side of the electrode, where the submerged arc would most likely jump off the electrode.

To contextualise the resistivity results presented in Figure 5a,b, they were firstly compared to high temperature bulk bed electrical resistivity values reported relatively recently for carbonaceous reductants (i.e., metallurgical coke, semi-coke, coal, and charcoal used in SAF smelting) [21]. Such high temperature bulk bed resistivities simulate the so-called coke bed (area around and below electrode tips in a SAF), where reductants serve as one of the primary electrical conductors [22]. At 1600 °C, it was found that the reductant bulk bed resistivities ranged from approximately 2×10^{-2} to 5×10^{-3} $\Omega{\cdot}$m [21]. These values were approximately 60 to 250 times higher (thus less conductive) than the approximate 8.40×10^{-5} $\Omega{\cdot}$m resistivity observed from 0.7 to 2.7 m below the contact shoes on the non-delta side of the electrode (reddish area on the right-hand side of Figure 5b). Correspondingly, the high temperature reductant bulk bed resistivities were approximately 2000 to 8000 times higher than the 2.4×10^{-6} $\Omega{\cdot}$m obtained at 2.7 m below the contact shoes on the electrode delta side (dark blue area in bottom left-hand corner of Figure 5b). Obviously, interparticle contact in a carbonaceous reductant bed is far inferior to the relatively solid carbonaceous mass of a Søderberg electrode, but the afore-mentioned comparison does indicate on a relative scale the conductivity ability of a Søderberg electrode.

To further contextualize the resistivity data presented in Figure 5a,b, it was compared to Søderberg electrode paste resistivity values derived from finite element thermoelectrical modelling [9,23]. These studies postulated resistivities to be 5.88×10^{-6} and 2.63×10^{-6} $\Omega{\cdot}$m in the electrode where the temperatures were 1000 and 2600 °C, respectively. Such temperatures correspond roughly to the 0.7 m (both delta and non-delta side) and 2.7 m (delta side only) positions below the contact shoes in Figure 5b, where resistivities were approximately 8.48×10^{-5} and 2.41×10^{-6} $\Omega{\cdot}$m, respectively. Although the modelled values [9,23] and values presented in the current work are in the same order, the differences in the corresponding values ($5.88 \times 10^{-6} - 2.63 \times 10^{-6} = 3.25 \times 10^{-6}$ $\Omega{\cdot}$m versus $8.48 \times 10^{-5} - 2.41 \times 10^{-6} = 8.24 \times 10^{-5}$ $\Omega{\cdot}$m) suggest that the models do not quickly enough account for lower resistivities as a function of electrode length below the contact shoes. Compared to the afore-mentioned models [9,23], a more comprehensive Søderberg electrode model has also previously been introduced [5]. This model might contain more detailed Søderberg electrode paste properties and might better account for the variation of such properties as a function of electrode length below the contact shoes. However, this model was essentially introduced as a black box to demonstrate the capabilities of a commercial company [5]; therefore, it was impossible to include the material properties specified therein in the current comparison. Furthermore, no evidence could be found that either of the three models [5,9,23] accounted for the very apparent asymmetrical variation of resistivities observed in the case study electrode (Figure 5b).

3.3. Degree of Graphitisation (DOG)

Figure 6a presents the average measured DOG and linear fitting of this data, as a function of length below the contact shoes and distance from the delta side of the electrode. Figure 6b presents the associated spatially interpolated 3D image. From both these figures (but particularly from Figure 6b) that visually represent the entire electrode cross section, it is evident that DOG is inversely related to the electrode resistivity (Figure 5b). This makes sense, since increased DOG of the electrode paste will result in lower resistivity (i.e., higher conductivity). It is particularly insightful to also note that only the electrode tip on the delta side of the electrode (red area in bottom left-hand corner of Figure 6b, which corresponded to a maximum DOG of approximately 0.84), approached full graphitisation. As previously stated (Section 2.3), DOG is unitless, with 1 indicating fully graphitised and 0 fully ungraphitised. The majority of the investigated cross-sectional electrode area had DOG ≤ 0.5. As far as the authors could assess, DOG for a real world or a modelled Søderberg electrode has not yet been presented in the peer reviewed public domain; therefore, it was not possible to contextualise the measured DOG with such data.

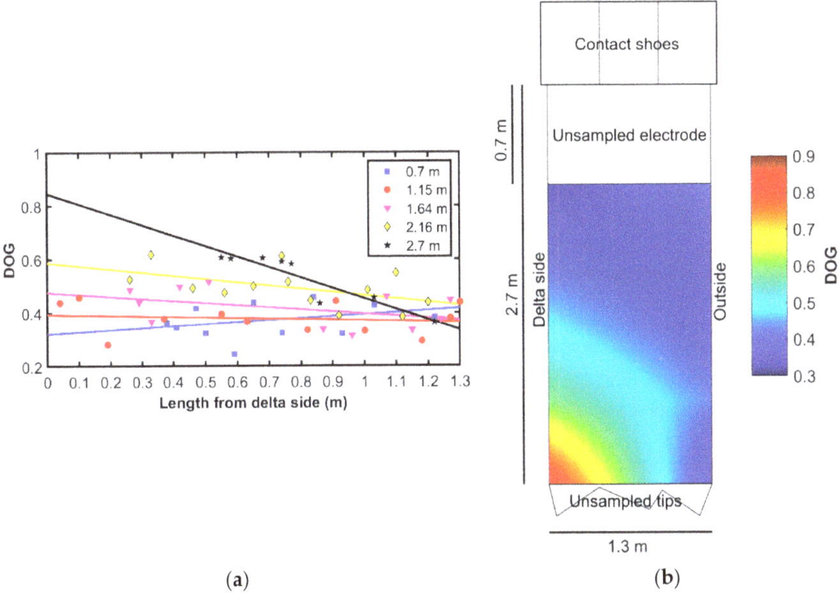

Figure 6. (**a**) Average measured degree of graphitisation (DOG) (markers) and linear fitting of this data (lines), as a function of length below the contact shoes and distance from the delta side of the electrode; (**b**) 3D rendering of the fitted line data associated with DOG at 10 mm intervals across the electrode at each of the lengths below the contacts shoes indicated in Figure 6a.

As indicated in the previous paragraph, to a large extent DOG determines the electrical resistivity of the Søderberg electrode. However, the DOG of the electrode paste is also related to other important characteristics. For instance, DOG is associated to possibly the most important aspect that cause electrode breaks, i.e., thermal stress. Particularly, green breaks (implying exposure of liquid paste in the break zone) in the contact shoes area are feared by operational personnel, due to the devastating consequences such breaks could have. In the most authoritative work on Søderberg electrode thermal stress, it is stated that some internal stresses are established during the process of electrode paste shrinkage below the contact shoes area, where the paste is baked at temperatures in excess of the basking isotherm temperature [5]. However, as previously stated, Søderberg electrode paste consists of a binder and a solid aggregate. In this case, the solid aggregate was calcined anthracite of various size factions, mixed to obtain the optimum composition. Although more environmentally friendly binders have been proposed than coal tar pitch, e.g., [24], it remains the most commonly used binder for Søderberg electrode pastes and was also used in the paste consumed by the case study electrode. In previous studies, it was proven that anthracite calcined at 1400 °C expands less than 1% and coal tar pitches prebaked at the maximum baking isotherm temperature (i.e., 475 °C, [11]) shrink approximately 12% if heated in a non-oxidising environment up to 1300 °C [4]. These opposing thermal expansion characteristics likely cause the thermal stresses below the contact shoes area referred to by Larsen et al. [5]. In contrast to calcined anthracite, uncalcined anthracite shrink 6 to 8% over the same temperature range. The significant difference between calcined and uncalcined anthracite clearly indicates how important the calcination process is to avoid additional thermal stresses forming just below the contact shoes area. Industrial calcination of anthracite to be used in electrode paste typically occur at temperatures above 1400 °C (as tested by Beukes et al. [4]); therefore, DOG similar to the blue area in Figure 6b (corresponding to DOG ≤ 0.5) is imparted on the electrode paste solid aggregate even before it is mixed into the electrode paste. The authors were not privileged to know the exact mixing ratio of the binder and aggregate used in the case study electrode. But, typically the mixture is dominated by aggregate, with the binder fractional contribution limited to just enable solid aggregate interparticle contact and proper paste flow characteristics above the softening point (to fill the casing in the area above the baking isotherm temperature).

3.4. Bulk Density

Figure 7a presents the average measured bulk density determined with ASTM C20-00 [19] (Section 2.4, 1st paragraph) as a function of length below the contact shoes and distance from the delta side of the electrode. Since ASTM C20-00 involves aqueous saturation of the samples via boiling, repeat analyses were not attempted. The density differences were relatively small (Figure 7a), therefore, even slight modification of the original characteristics that might occur during analysis or re-drying of the samples in preparation for reanalyses could make such reanalyses results misleading. Additionally, larger cylindrical sample pellets were used in this specific method (Section 2.4, 1st paragraph), which implied that significant additional sample material would have been required for reanalyses. This was not feasible, since a finite amount of material was available after sampling (Section 2.1). As is evident from the ASTM C20-00 results (Figure 7a), bulk densities did not indicate significant differences as a function of length below the contact shoes. Therefore, a single linear fitted line was drawn through all the data, which indicated a slight increase in densities toward the delta side of the electrode.

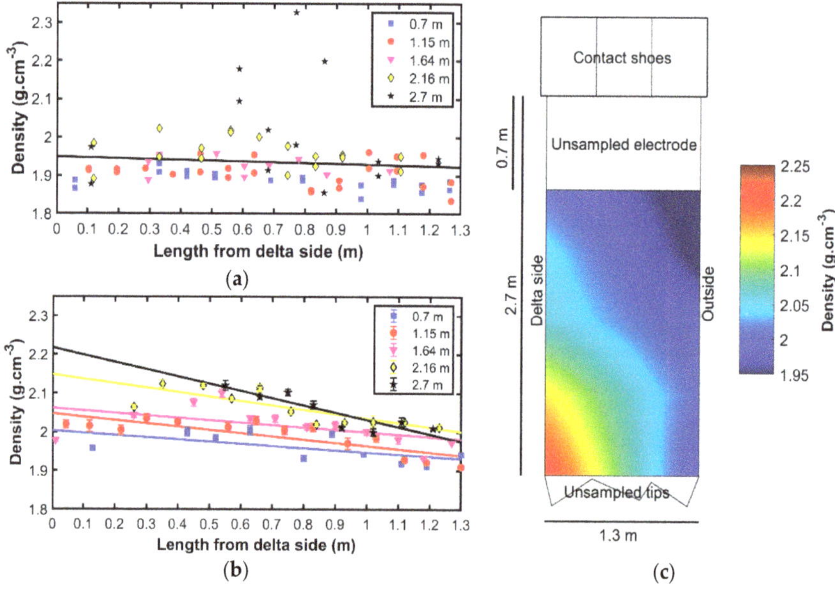

Figure 7. (**a**) ASTM C20-00 average measured bulk densities (markers) as a function of length below the contact shoes and distance from the delta side of the electrode. A single linear fitting of all this data (line) is also indicated; (**b**) gas pycnometry average measured bulk densities (markers) and linear fitting of this data (lines), as a function of length below the contact shoes and distance from the delta side of the electrode; (**c**) 3D rendering of the fitted line data associated with bulk densities at 10 mm intervals across the electrode at each of the lengths below the contacts shoes indicated in Figure 7b.

To verify that densities did not differ as a function of length below the contact shoes and to determine standard deviations, bulk densities were also determined with gas pycnometry (Section 2.4, 2nd paragraph). The gas pycnometry densities (Figure 7b), in contrast to the ASTM C20-00 results (Figure 7a), did indicate that bulk densities differed as a function of length below the contact shoes. This indicates that gas pycnometry is possibly a more precise method than ASTM C20-00, or alternatively that the difference in atom/molecule size (i.e., He used in gas pycnometry vs. H_2O used ASTM C20-00) play a significant role in the measured results. To visualise these gas pycnometry results, Figure 7c presents the associated spatially interpolated 3D image, based on the fitted data presented in Figure 7b. Considering the very similar perspectives indicated by the bulk densities presented here (Figure 7c) and the DOG presented earlier (Figure 6b), it is inferred that the increased bulk densities observed, particularly on the delta side of the electrode as a function of length below the contact shoes, were due to increased DOG. Increased DOG will result in further shrinkage of the carbonaceous Søderberg electrode paste material.

3.5. Porosity

Figure 8a presents the average measured porosity. As is evident, no significant difference in porosity as a function of length below the contact shoes were observed. However, a slight increase in porosity was observed, near the electrode perimeter on both the delta and non-delta side. Therefore, a 2nd order polynomial (solid line in Figure 8a) fitted the data best. The corresponding spatially interpolated 3D image is presented in Figure 8b.

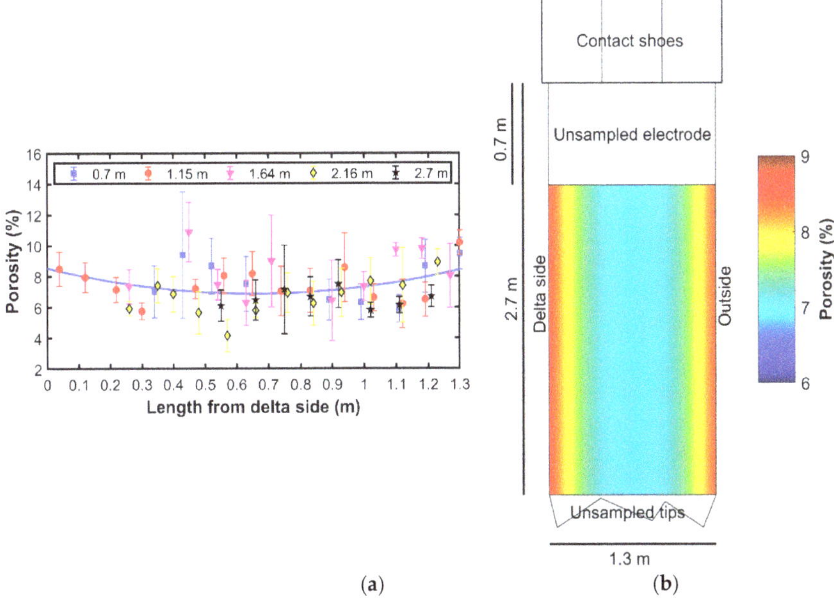

Figure 8. (**a**) Average measured porosity (markers), as a function of length below the contact shoes and distance from the delta side of the electrode, as well as a single 2nd order polynomial fit of this data (line); (**b**) 3D rendering of the fitted line data associated with porosities at 10 mm intervals across the electrode at each of the lengths below the contacts shoes indicated in Figure 8a.

Since the industrial application of the Søderberg electrode started, it has been proven that volatiles released from the liquid paste just above the baking isotherm (indicated by black dotted line in Figure 9a) cannot rise if proper (high enough) liquid and solid paste levels are maintained e.g., [25]. Conversely, if "smoke" is observed at the top of the electrode column casing, it serves as a serious warning signal to operational personal that the liquid paste level of that electrode is critically low and/or that the liquid paste is too hot. Since the volatiles just above the baking isotherm cannot rise, they escape downward through the pores of the already baked electrode [25]. Cracking of these descending volatiles (due to the high temperature of the baked electrode) leaves a considerable amount of carbon in the pores of the baked electrode paste just below the contact shoes area, which improves its quality [25]. This process is illustrated by the purple arrows in the enlarged area of Figure 9a. The blue temperature isotherm in the enlarged area of Figure 9b represents a slightly hotter isotherm temperature than the baking isotherm. Considering this diagram, one would assume that the route of least resistance for volatiles to escape would be towards the outside of the electrode (i.e., shortest route of baked electrode paste), not the electrode core. Theoretically, this process would therefore slightly reduce the porosity of the electrode towards its perimeter, which is contrary to the experimental results presented (Figure 8a,b).

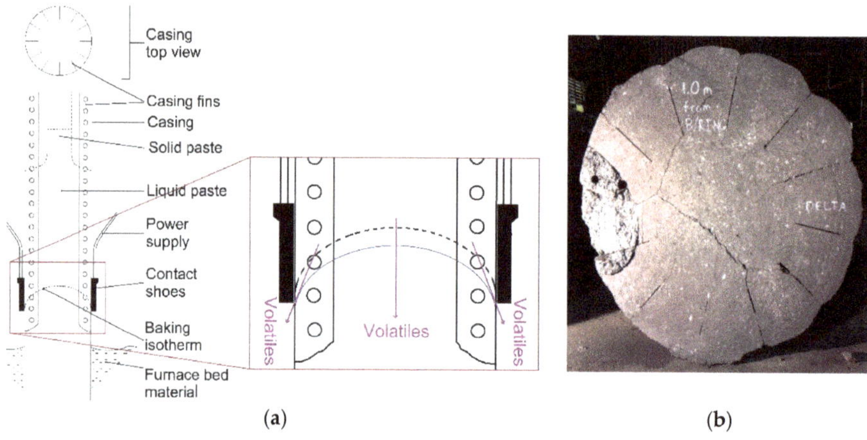

(a)   (b)

Figure 9. (**a**) Representation of a partial Søderberg electrode column, adapted from Beukes et al. [4], with the contact shoes area enlarged (indicated by the square red block); (**b**) cross section of another electrode from the same SAF that was cut off with a diamond wire saw approximately 1.0 m below the contact shoes.

There could be two other factors that counter the possible decrease in porosity towards the electrode perimeter, caused by descending volatile cracking. Firstly, liquid electrode paste segregation (i.e., partial separation of the pitch binder and solid aggregate), as illustrated by Nelson and Prins [8], could result in significant differences in the baked electrode properties considered diagonally across an electrode. However, calcined anthracite has a much higher DOG than coal tar pitch baked above the maximum baking isotherm temperature (i.e., 475 °C, [11]). Therefore, if paste segregation caused the differences in observed porosities (Figure 8b), corresponding differences in DOG diagonally across the case study electrode would likely also have evident at the first level below the contact shoes investigated (i.e., 0.7 m). This was not the case (Figure 6b); therefore, paste segregation is unlikely to be the cause. Also, visual inspection of the sampled electrode did not indicate any segregation. The second possible contributing factor that could increase porosity of the case study electrode is chemical corrosion, i.e., oxidation of the electrode carbon above the bed material. As previously stated, this can especially occur in an open-SAF, as considered here [20]. Figure 9b shows the diagonal cross section of another electrode from the same SAF that was cut off approximately 1.0 m below the contacts shoes. As is evident from this electrode, corrosion on the outer perimeter was more significant in the areas where the casing fins occurred. This indicates that the gaps remaining after the casing burned off and/or melted away, formed paths for oxygen penetration, and associated oxidation of the carbon material. This describes the process on a macroscale, but it is postulated that the same process occurs on a pore size scale, which will increase pore volumes and hence porosity preferentially on the electrode perimeter.

3.6. Compressive Breaking Strength

Figure 10 presents the average measured compressive breaking strength as a function of length below the contact shoes and distance from the electrode delta side. As is evident, no significant difference was observed over the entire electrode area investigated. This indicates that compressive breaking strength was not related to DOG, as was the case for electrical resistivity and bulk density, nor was it influenced by physical and/or chemical processes, as postulated for the porosities.

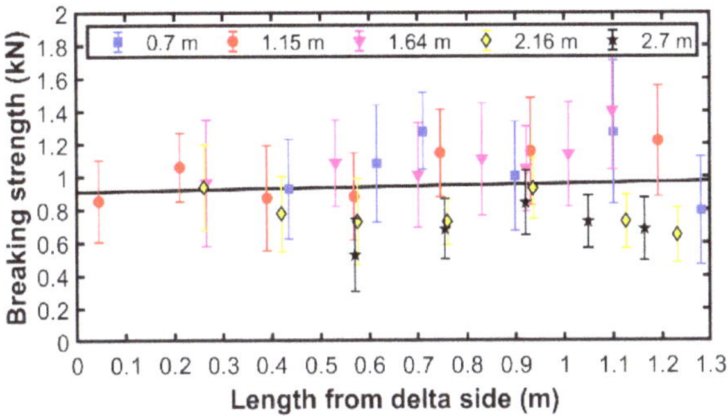

Figure 10. Average measured compressive breaking strength (markers), as a function of length below the contact shoes and distance from the delta side of the electrode, as well as a single linear fit of this data (line).

4. Conclusions

As far as the authors could assess, this work presents the first characterisation of a real-world Søderberg electrode, presented in a peer reviewed journal in the public domain. Obviously, a study such as this has limitations. For instance, measurements of the characteristics were conducted at room temperature, instead of at the operational temperatures. Also, the study was conducted on an electrode removed for an open-SAF producing ferrochrome, for which some characteristics might differ from Søderberg electrodes operating in other furnace types and smelting different materials. However, all these limitations are similar to furnace dig-out studies, e.g., [26–28], which have to a large degree formed our current understanding of zones within furnaces. Another data limitation of the current work was that the case study electrode did not include the area directly above, in, and just below the contact shoes. Understanding this area is critical. Unfortunately, this section of the electrode could not be obtained, since that would have entailed starting up the furnace with a green electrode tip.

The results presented in this paper provided valuable insights, but also indicated questions/topics that need to be considered in future research. The operational data for the month prior to the furnace rebuild, during which the case study electrode was removed, proved that the electrode was a representative specimen. The characterisation results indicated no significant electrical resistivity, DOG, and bulk density changes from 0.7 to 2.7 m on the non-delta side of the electrode, while these characteristics changed relatively significantly on the delta side of the electrode. The area where the submerged arc would mostly like jump off the electrode clearly had the lowest resistivity, as well as highest DOG and bulk density. It is currently not evident from literature if these very asymmetrical characteristics can be accurately predicted by Søderberg electrode models, e.g., [5,9,23]. However, all such models predict temperature isotherms/curves. The temperature to which Søderberg electrode paste is exposed determines the DOG, to a large degree. As shown here, the DOG in turn determine to a large extent, the resistivities and bulk densities. Therefore, it would be advantageous if Søderberg electrode models could accurately replicate the DOG presented here. To aid such comparison, the resistivities, DOG, and bulk densities associated with the fitted lines at 10 mm intervals across the diameter of the electrode at each of the investigated lengths below the contacts shoes prior to being spatial interpolated to generate the 3D images are presented in Appendix A. For future experimental studies, it is recommended that the DOG of calcined anthracite and coal tar pitch be investigated separately (not as a mixture in electrode paste), in relation to electrical resistivity and bulk density, as well as other characteristics such as thermal expansion. This will enable better understanding of the individual contribution of these two components, to the combined results observed for electrode paste.

The determination of porosity with CT scanning combined with image processing is likely more accurate than conventional methods. However, the manual marking of the areas to be included in the calculations for porosity made the method time consuming and less attractive. Research has already been initiated to automate the determination of the particle rims within CT image slices and calculation of porosities and pore volumes. From the presented porosity data, no significant difference in porosities, as a function of length below the contact shoes, were evident. However, slight increases in porosities were observed near the electrode perimeter on both the delta and non-delta sides, which was fitted best with a 2nd order polynomial function. It was postulated that chemical corrosion, i.e., oxidation of the carbonaceous material, increases the pore volumes in the baked electrode paste near the perimeter.

The compressive breaking strength data revealed no significant trends. Therefore, operational personnel should not be concerned about breaking strength in the region well below the baking isotherm being during stable operation. As previously stated [5], thermal stresses that form in and just below the baking isotherm, as well as thermal stress differences between the perimeter and core of the electrode formed due to prolong furnace downtime, remain the most likely reasons for breaks of electrodes with no other apparent dysfunctionalities.

Considering the value of studies such as this, hopefully the current work will also encourage the donation of similar case study electrodes to researchers. As an example, the authors belatedly became aware of 2 separate 63 MVA closed-SAFs, for which the electrode outer casing diameters were increased from 1.5 to 1.6 m. In order to do so, 6 entire electrode columns, including the casings with un-melted solid paste at the top, down to the baked electrode tips at the bottom, were removed and discarded. However, no one in the research community was notified, nor did the company/operational personnel reach out to researchers. Such large scale, real-world samples are scarce; therefore, opportunities such as this should not be missed.

Author Contributions: Conceptualisation, J.P.B., P.G.v.Z., and D.D.; measurement methodology, J.P.B., R.I.G., M.T., E.R., and J.D.S.; experimental investigation, R.I.G. and M.M.; writing—original draft preparation, J.P.B.; writing—review and editing, all co-authors; study supervision, J.P.B. and P.G.v.Z.; funding acquisition, J.P.B., P.G.v.Z., M.T., and E.R. All authors have read and agreed to the published version of the manuscript.

Funding: Funding for this project from SFI Metal Production (Centre for Research-based Innovation, 237738) and the Thanos project (INPART project 309475), both supported by the Research Council of Norway (RCN), is acknowledged, as well as industrial partners that support SFI Metal Production.

Acknowledgments: The authors thank the Wonderkop Smelter, which is part of Glencore Alloys, South Africa, for donating the Søderberg electrode considered in this study and supplying the operational data for the electrode.

Conflicts of Interest: The authors declare no conflict of interest. The work described is original research that has not been published previously, and is not under consideration for publication elsewhere, in whole or in part.

Appendix A

Table A1. The electrical resistivity, DOG, and bulk density data associated with the fitted lines in Figure 5a, Figure 6a, and Figure 7b, at 10 mm intervals across the diameter of the electrode at each of the investigated lengths below the contacts shoes, respectively. This data was spatially interpolated to generate the 3D images presented in Figure 5b, Figure 6b, and Figure 7c, respectively.

Distance from Delta Side (m)	Distance below the Contact Shoes (m)				
	0.70	1.15	1.64	2.16	2.70
Resistivity ($\Omega \cdot$m)					
0	7.42×10^{-5}	7.34×10^{-5}	4.66×10^{-5}	2.92×10^{-5}	2.41×10^{-6}
0.1	7.51×10^{-5}	7.37×10^{-5}	4.98×10^{-5}	3.37×10^{-5}	9.25×10^{-6}
0.2	7.59×10^{-5}	7.41×10^{-5}	5.30×10^{-5}	3.82×10^{-5}	1.61×10^{-5}
0.3	7.67×10^{-5}	7.45×10^{-5}	5.62×10^{-5}	4.27×10^{-5}	2.29×10^{-5}
0.4	7.75×10^{-5}	7.49×10^{-5}	5.94×10^{-5}	4.72×10^{-5}	2.98×10^{-5}
0.5	7.83×10^{-5}	7.52×10^{-5}	6.26×10^{-5}	5.17×10^{-5}	3.66×10^{-5}
0.6	7.91×10^{-5}	7.56×10^{-5}	6.58×10^{-5}	5.62×10^{-5}	4.35×10^{-5}
0.7	7.99×10^{-5}	7.60×10^{-5}	6.91×10^{-5}	6.07×10^{-5}	5.03×10^{-5}
0.8	8.07×10^{-5}	7.64×10^{-5}	7.23×10^{-5}	6.52×10^{-5}	5.71×10^{-5}
0.9	8.15×10^{-5}	7.68×10^{-5}	7.55×10^{-5}	6.97×10^{-5}	6.40×10^{-5}
1.0	8.24×10^{-5}	7.71×10^{-5}	7.87×10^{-5}	7.42×10^{-5}	7.08×10^{-5}
1.1	8.32×10^{-5}	7.75×10^{-5}	8.19×10^{-5}	7.88×10^{-5}	7.77×10^{-5}
1.2	8.40×10^{-5}	7.79×10^{-5}	8.51×10^{-5}	8.33×10^{-5}	8.45×10^{-5}
1.3	8.48×10^{-5}	7.83×10^{-5}	8.83×10^{-5}	8.78×10^{-5}	9.13×10^{-5}
DOG					
0	0.32	0.39	0.48	0.59	0.86
0.1	0.33	0.39	0.47	0.58	0.81
0.2	0.34	0.39	0.46	0.56	0.77
0.3	0.34	0.39	0.45	0.55	0.73
0.4	0.35	0.39	0.45	0.54	0.69
0.5	0.36	0.38	0.44	0.53	0.65
0.6	0.37	0.38	0.43	0.52	0.61
0.7	0.37	0.38	0.42	0.50	0.57
0.8	0.38	0.38	0.41	0.49	0.53
0.9	0.39	0.38	0.41	0.48	0.49
1.0	0.40	0.37	0.40	0.47	0.45
1.1	0.40	0.37	0.39	0.46	0.41
1.2	0.41	0.37	0.38	0.44	0.38
1.3	0.42	0.37	0.37	0.43	0.34
Bulk density (g$\cdot$cm^{-3})					
0	2.00	2.05	2.06	2.15	2.22
0.1	1.99	2.04	2.06	2.14	2.20
0.2	1.99	2.03	2.05	2.13	2.18
0.3	1.99	2.02	2.04	2.11	2.16
0.4	1.98	2.01	2.04	2.10	2.14
0.5	1.98	2.01	2.03	2.10	2.12
0.6	1.97	2.00	2.02	2.08	2.11
0.7	1.96	1.99	2.01	2.07	2.09
0.8	1.96	1.98	2.01	2.06	2.07
0.9	1.95	1.97	2.00	2.05	2.05
1.0	1.95	1.96	2.00	2.04	2.03
1.1	1.94	1.96	1.99	2.02	2.01
1.2	1.94	1.95	1.99	2.01	2.00
1.3	1.93	1.94	1.98	2.00	1.98

References

1. Richards, J.W. The Söderberg self-Baking continuous electrode. In Proceedings of the Thirty-Seventh General Meeting of the American Electrochemical Society, Boston, MA, USA, 8 April 1920; pp. 169–188.

2. Larsen, B. Electrode models for Søderberg electrodes. In Proceedings of the Fifth Platinum Conference, Sun City, South Africa, 18 September 2012; pp. 275–286.

3. Skjeldestad, A.; Tangstad, M.; Lindstad, L.; Larsen, B. Temperature profiles in Søderberg electrodes. In Proceedings of the Fourteenth International Ferroalloys Congress (INFACON), Kiev, Ukraine, 31 May–4 June 2015; pp. 327–337.

4. Beukes, J.P.; Roos, H.; Shoko, L.; Van Zyl, P.G.; Neomagus, H.W.J.P.; Strydom, C.A.; Dawson, N.F. The use of thermomechanical analysis to characterise Söderberg electrode paste raw materials. *Miner. Eng.* **2013**, *46*, 167–176. [CrossRef]

5. Larsen, B.; Feldborg, H.; Halvorsen, S.A. Minimizing thermal stress during shutdown of Søderberg electrodes. In Proceedings of the Thirteenth International Ferroalloys Congress (INFACON), Almaty, Kazakhstan, 9–13 June 2013; pp. 453–466.

6. Meyjes, R.P.; Venter, J.; Van Rooyen, U. Advanced modelling and baking of electrodes. In Proceedings of the Twelfth International Ferroalloys Congress (INFACON), Helsinki, Finland, 6–9 June 2010; pp. 779–788.

7. Ray, C.R.; Sahoo, P.K.; Rao, S.S. Electrode management–investigation into soft breaks at 48MVA FeCr closed furnace. In Proceedings of the Eleventh International Ferroalloys Congress (INFACON), New Delhi, India, 18–21 February 2007; pp. 742–751.

8. Nelson, L.R.; Prins, F.X. Insights into the influences of paste addition and levels on Söderberg electrode management. In Proceedings of the Tenth International Ferroalloys Congress (INFACON), Cape Town, South Africa, 1–4 February 2004; pp. 418–431.

9. Mc Dougall, I.; Smith, C.F.R.; Olmstead, B.; Gericke, W.A. Finite element model of a Søderberg electrode with an application in casing design. In Proceedings of the Tenth International Ferroalloys Congress (INFACON), Cape Town, South Africa, 1–4 February 2004; pp. 575–584.

10. Larsen, B.; Amaro, J.P.M.; Nascimento, S.Z.; Fidje, K.; Gran, H. Melting and densification of electrode paste briquettes in Søderberg electrodes. In Proceedings of the Tenth International Ferroalloys Congress (INFACON), Cape Town, South Africa, 1–4 February 2004; pp. 405–417.

11. Shoko, L.; Beukes, J.P.; Strydom, C.A. Determining the baking isotherm temperature of Söderberg electrodes and associated structural changes. *Miner. Eng.* **2013**, *49*, 33–39. [CrossRef]

12. White, J.F.; Rigas, K.; Andersson, S.P.; Glaser, B. Thermal properties of Söderberg electrode materials. *Metall. Mater. Trans. B* **2020**, *51*, 1928–1932. [CrossRef]

13. Karalis, K.T.; Karkalos, N.; Cheimarios, N.; Antipas, G.S.E.; Xenidis, A.; Boudouvis, A.G. A CFD analysis of slag properties, electrode shape and immersion depth effects on electric submerged arc furnace heating in ferronickel processing. *Appl. Math. Model.* **2016**, *40*, 9052–9066. [CrossRef]

14. Shoko, L.; Beukes, J.P.; Strydom, C.A.; Larsen, B.; Lindstad, L. Predicting the toluene-And quinoline insoluble contents of coal tar pitches used as binders in Söderberg electrodes. *Int. J. Miner. Process.* **2015**, *144*, 46–49. [CrossRef]

15. Jordan, D.T.; Hockaday, C.J. Development of a camera-Based Söderberg electrode slip measurement device. In Proceedings of the Thirteenth International Ferroalloys Congress (INFACON), Almaty, Kazakhstan, 9–13 June 2013; pp. 417–426.

16. Heikkilä, A.M.; Pussinen, J.H.; Mattila, O.J.; Fabritius, T.M.J. About electrical properties of chromite pellets –Effect of reduction degree. *Steel Res. Int.* **2015**, *86*, 121–128. [CrossRef]

17. Zou, L.; Huang, B.; Huang, Y.; Huang, Q.; Wang, C. An investigation of heterogeneity of the degree of graphitization in carbon-carbon composites. *Mater. Chem. Phys.* **2003**, *82*, 654–662. [CrossRef]

18. Zhang, Z.; Wang, Q. The new method of XRD measurement of the degree of disorder for anode coke material. *Crystals* **2017**, *7*, 1–10. [CrossRef]

19. C20-00, ASTM. *Standard Test Method for Apparent Porosity, Water Absorption, Apparent Specific Gravity, and Bulk Density of Burned Refractory Brick and Shapes by Boiling Water*; ASTM International: West Conshohocken, PA, USA, 2010.

20. Beukes, J.P.; Du Preez, S.P.; Van Zyl, P.G.; Paktunc, D.; Fabritius, T.; Päätalo, M.; Cramer, M. Review of Cr(VI) environmental practices in the chromite mining and smelting industry–Relevance to development of the Ring of Fire, Canada. *J. Clean. Prod.* **2017**, *165*, 874–889. [CrossRef]

21. Surup, G.; Pedersen, T.; Chaldien, A.; Beukes, J.P.; Tangstad, M. Electrical resistivity of carbonaceous bed material at high temperature. *Processes* **2020**, *8*, 0933. [CrossRef]

22. Tangstad, M.; Beukes, J.P.; Steenkamp, J.; Ringdalen, E. Chapter 14. Coal-Based reducing agents in ferroalloys and silicon production. In *New Trends in Coal Conversion*; Suárez-Ruiz, I., Rubiera, F., Diez, M., Eds.; Elsevier: Duxford, UK, 2019; pp. 405–438. Available online: http://dx.doi.org/10.1016/B978-0-08-102201-6.00014-5 (accessed on 30 October 2020).

23. Bermúdez, A.; Bullón, J.; Pena, F. A finite element method for the thermoelectrical modelling of electrodes. *Commun. Numer. Methods Eng.* **1998**, *14*, 581–593. [CrossRef]

24. Ciesińska, W.; Zieliński, J.; Brzozowska, T. Thermal treatment of pitch-polymer blends. *J. Therm. Anal. Calorim.* **2009**, *95*, 193–196. [CrossRef]

25. Sem, M.O. The Söderberg self-baking electrode. *J. Electrochem. Soc.* **1954**, *101*, 487–492. [CrossRef]

26. Wedepohl, A.; Barcza, N.A. The "dig-Out" of a ferrochromium furnace. *Spec. Publ. Geol. Soc. S. Afr.* **1983**, *7*, 351–363.

27. Nell, J.; Joubert, C. Phase chemistry of digout samples from a ferrosilicon furnace. In Proceedings of the Thirteenth International Ferroalloys Congress (INFACON), Almaty, Kazakhstan, 9–13 June 2013; pp. 265–272.

28. Steenkamp, J.D.; Gous, J.P.; Grote, W.; Cromarty, R.; Gous, H.J. Process zones observed in a 48 MVA submerged arc furnace producing silicomanganese according to the ore-Based process. In Proceedings of the Extraction 2018, The Minerals, Metals & Materials Series, Ottawa, ON, Canada, 26–29 August 2018. Available online: https://doi.org/10.1007/978-3-319-95022-8_50 (accessed on 30 October 2020).

Article

Production of Ferronickel Concentrate from Low-Grade Nickel Laterite Ore by Non-Melting Reduction Magnetic Separation Process

Guorui Qu [1,2], **Shiwei Zhou** [2,3], **Huiyao Wang** [2,3], **Bo Li** [1,2,3] and **Yonggang Wei** [1,2,3,*]

[1] State Key Laboratory of Complex Nonferrous Metal Resources Clean Utilization, Kunming University of Science and Technology, Kunming 650093, China; qugr11@126.com (G.Q.); libokmust@163.com (B.L.)

[2] Engineering Research Center of Metallurgical Energy Conservation and Emission Reduction, Ministry of Education, Kunming University of Science and Technology, Kunming 650093, China; swzhou90@126.com (S.Z.); wanghuiyao940921@163.com (H.W.)

[3] Faculty of Metallurgical and Energy Engineering, Kunming University of Science and Technology, Kunming 650093, China

* Correspondence: weiygcp@aliyun.com

Received: 20 November 2019; Accepted: 9 December 2019; Published: 12 December 2019

Abstract: The production of ferronickel concentrate from low-grade nickel laterite ore containing 1.31% nickel (Ni) was studied by the non-melting reduction magnetic separation process. The sodium chloride was used as additive and coal as a reductant. The effects of roasting temperature, roasting duration, reductant dosage, additive dosage, and grinding time on the grade and recovery were investigated. The optimal reduction conditions are a roasting temperature of 1250 °C, roasting duration of 80 min, reductant dosage of 10%, additive dosage of 5%, and a grinding time of 12 min. The grades of nickel and iron are improved from 2.13% and 51.12% to 8.15% and 64.28%, and the recovery of nickel is improved from 75.40% to 97.76%. The research results show that the additive in favor of the phase changes from lizardite phase to forsterite phase. The additive promotes agglomeration and separation of nickel and iron.

Keywords: nickel laterite; non-melting reducing; sodium chloride; magnetic separation

1. Introduction

As an important metal, nickel has an important position in industrial production due to its good physicochemical properties [1,2]. Nickel resources mainly exist in the form of nickel sulfide ore and nickel laterite ore. Among them, nickel laterite ore accounts for about 70% of the total nickel reserves in the world [3]. Due to the continuous decline of global nickel sulfide ore reserves, the development and utilization of nickel laterite ore have received increasing attention.

The treatment method of laterite ore can be divided into two types: hydrometallurgy and pyrometallurgy. The pyrometallurgical process has high requirements for the Ni grade, so it is impossible to treat the low-grade nickel laterite [4,5]. At the same time, high temperatures will cause slag and metal rings to form, which is not conducive to the production of ferronickel [6]. The disadvantage of the wet process is that the process is long and the wastewater is difficult to handle [3,7,8]. Therefore, the use of the "low-temperature roast reduction-magnetic separation process" to deal with such resources is attracting everyone's attention [6,9].

The roasted reduction-magnetic separation process is mainly used for the treatment of laterite ore, which requires the laterite ore to be crushed and ground. Then the reductant and additive are mixed with the raw material and roasted at a certain temperature. After that, the roasted products were ground and wet-type magnetic separation was carried out to obtain ferronickel concentrate.

In low-grade nickel laterite ore, the presence of nickel is extremely complex. Therefore, the addition of additives is often used to treat such laterites ore. There are generally two types of additives, one is sulfur and sulfate, and the other is chloride. Valix et al. [10] showed that the addition of elemental S inhibited the formation of forsterite in the reduction process, thus achieving selective separation of nickel and iron. Harris et al. [11] studied the selective vulcanization of nickel laterite at low temperatures, and the results show that nickel oxide can be selectively vulcanized. Although the addition of sulfur and sulfate can contribute to the recovery of nickel, the S content in the product is high [12]. Chloride is characterized by high reactivity of chlorine and the low melting point of metal chloride, which can selectively separate valuable metals [13,14]. Okamoto et al. [15] have shown that with the use of chloride salts as additives to treat nickel laterite, the chlorination reaction occurring during the roasting process can chlorinate and reduce the valuable metals in the ore. Ilic et al. [16,17] studied the selective extraction of nickel from raw materials containing iron and nickel by adding calcium chloride. The results show that it is feasible to selectively extract nickel by adding calcium chloride. Li et al. [18] used calcium chloride as an additive to reduce nickel laterite ore. The results show that calcium chloride can promote the migration of FeO to destroy the structure of the fayalite, thereby increasing the activity of nickel. Fan et al. [19] used ferrous chloride tetrahydrate as a chlorinating agent to selectively chlorinate pre-reduced nickel sulfide laterite, which achieved good results. Compared with other chlorides, sodium chloride has the advantages of wide sources and low prices. In this paper, the non-melting reduction-magnetic separation process of low-grade nickel laterite ore containing 1.13% of nickel was carried out with sodium chloride as an additive. The selective reduction of nickel is achieved by controlling the roasting temperature, the roasting duration, the reductant dosage, additive dosage, and the grinding time. The result is a concentrate with a high nickel grade and high recovery. In addition, the phase change of the ore under the action of the additive was studied.

2. Materials and Methods

2.1. Analysis of Nickel Laterite

The X-ray diffraction (XRD) analysis and chemical analysis of the nickel laterite ore sample is shown in Figure 1 and Table 1. The results in Table 1 show that the nickel content in the nickel laterite ore sample is only 1.13%, and the iron content is 35.79%. The XRD (Rigaku Corporation, Tokyo, Japan) results in Figure 1 show that the main phases in the sample are lizardite ($Mg_3[(Si,Fe)_2O_5](OH)_4$), goethite (FeO(OH)), and maghemite (γ-Fe_2O_3). The nickel is usually found in serpentine. Due to the low nickel content, the diffraction peaks of the corresponding nickel-containing phases could not be detected.

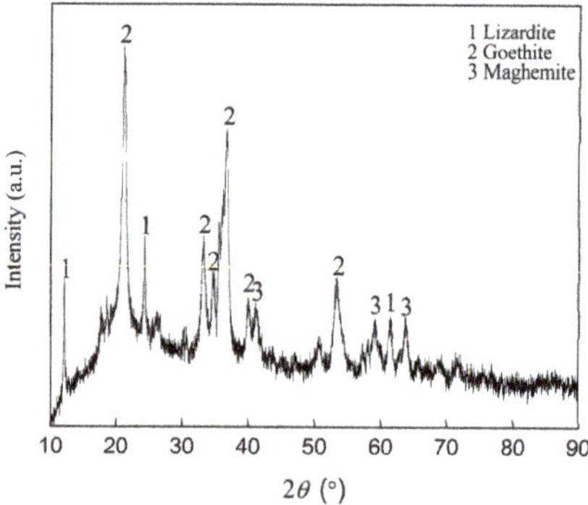

Figure 1. XRD pattern of nickel laterite ore.

Table 1. Chemical analysis of the nickel laterite ore sample.

Component	TFe(total)	Ni	Co	MgO	SiO$_2$	Al$_2$O$_3$
Content weight (wt. %)	35.79	1.13	0.097	3.65	10.38	10.31

2.2. Reductants and Additives

The experiment used anthracite as a reductant and analyzed pure sodium chloride as an additive. Industrial analysis of anthracite in Table 2 shows a fixed carbon content of 76.43%

Table 2. Industry analysis of the reductant.

Component	Fixed Carbon	Volatile Matter	Ash	Moisture
Content wt. %	76.43	7.78	15.29	1.02

2.3. Methods

The 10 g nickel laterite ore sample is crushed and ground to below 250 μm, mixed with anthracite and additives according to their certain mass ratio, and put into a crucible. The crucible was then placed in a tube furnace and roasted under a nitrogen atmosphere. After the reduction, the furnace was cooled to room temperature. The roasted product was ground and wet magnetic separation was able to be used in order to obtain the concentrate. The experimental procedure and detailed experimental conditions are shown in Figure 2 and Table 3. Roasting reduction equipment GSL-1500X tube type high-temperature sintering furnace (Hefei Ke Jing Materials Technology CO.,LTD, Hefei, China) as well as XZM-100 vibration grinding machine (Wuhan Exploration Machinery Factory, Wuhan, China) and DTCXG-ZN50 type magnetic separation tube (Tangshan DT Electrical Equipment CO.,LTD, Tangshan, China) (magnetic field strength 200 mT) were used. The recoveries of Ni and Fe were calculated by the following formula:

$$X = \frac{m \cdot w}{M \cdot W} \cdot 100\%,$$

where X is the recovery of nickel or iron, m is the weight of concentrate, w is the nickel or iron content in the concentrates, M is the weight of raw ore, and W is the nickel or iron content in the raw ore.

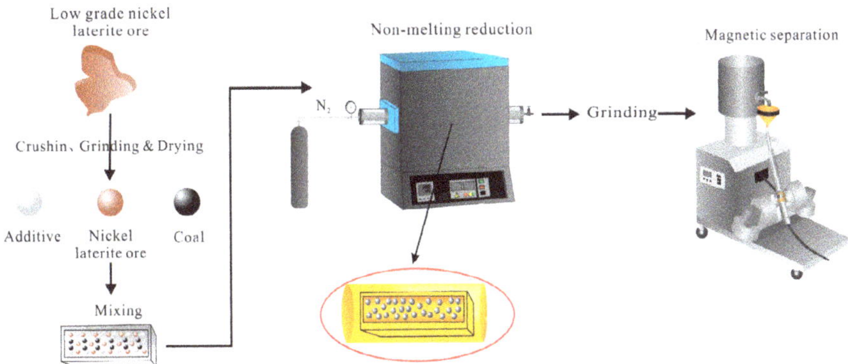

Figure 2. Experimental flowsheet.

Table 3. List of process parameters studied during experiments.

Studied Parameters	Range
Roasting temperature/°C	900, 1000, 1100, 1200, 1250, 1300
Roasting duration/min	10, 30, 60, 80, 100
Reductant dosage/wt. %	2, 5, 10, 13, 15
NaCl dosage/wt. %	0, 5, 8, 10, 15
Grinding time/min	2, 4, 8, 12, 16

3. Results and Discussion

3.1. Thermodynamic Analysis of Selective Reduction

First, sodium chloride is hydrolyzed to generate hydrogen chloride gas. Second, the hydrogen chloride gas can chlorinate valuable metals in ore. In addition, iron oxide is difficult to be reduced by hydrogen chloride. Therefore, iron oxide is first reduced to FeO and then chlorinated. The relevant equations are as follows:

$$2NaCl(s) + SiO_2(s) + H_2O(g) = Na_2O \cdot SiO_2(s) + 2HCl(g), \tag{1}$$

$$3Fe_2O_3(s) + C(s) = 2Fe_3O_4(s) + CO(g), \tag{2}$$

$$Fe_3O_4(s) + CO(g) = 3FeO(s) + CO_2(g), \tag{3}$$

$$FeO(s) + 2HCl(g) = FeCl_2(s) + H_2O(g), \tag{4}$$

$$NiO(s) + 2HCl(g) = NiCl(s) + H_2O(g). \tag{5}$$

The fixed carbon is hydrolyzed at a high temperature to form hydrogen and carbon monoxide or carbon dioxide. The metal chloride is reduced by these educing gases. The relevant equations are as follows:

$$C(s) + H_2O(g) = H_2(g) + CO(g), \tag{6}$$

$$C(s) + 2H_2O(g) = 2H_2(g) + CO_2(g), \tag{7}$$

$$NiCl_2(s) + H_2(g) = Ni(s) + 2HCl(g), \tag{8}$$

$$NiCl_2(s) + CO(g) + H_2O(g) = Ni(s) + 2HCl(g) + CO_2(g), \tag{9}$$

$$FeCl_2(s) + H_2(g) = Fe(s) + 2HCl(g), \tag{10}$$

$$FeCl_2(s) + CO(g) + H_2O(g) = Fe(s) + 2HCl(g) + CO_2(g). \tag{11}$$

The Gibbs free energy of the (8)–(11) reaction formula was calculated using the HSC Chemistry 6.0 software (Outotec Oyj, Espoo, Finland) as shown in Figure 3. As can be seen from the figure, the temperatures at which $NiCl_2$ is reduced by hydrogen and carbon monoxide are 435 °C and 335 °C. The temperatures at which $FeCl_2$ is reduced by hydrogen and carbon monoxide are 944 °C and 1022 °C. The above mentioned temperature only represents the reaction spontaneous temperature. The higher the temperature, the more easily the chemical reaction is to occur, the more sufficient the reaction. Compared to the temperature at which $FeCl_2$ was reduced, the temperature at which $NiCl_2$ was reduced is extremely low. Therefore, selective reduction can be achieved by controlling the temperature.

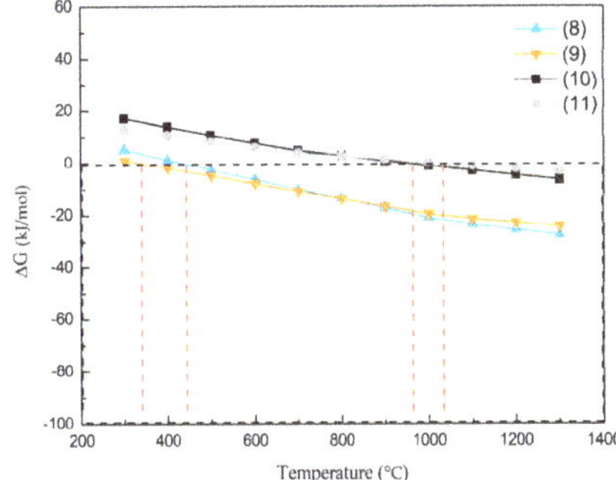

Figure 3. Gibbs free energy change curves of reactions with the change of temperature.

3.2. Effect of Different Influencing Factors on the Grade and Recovery of Ni and Fe

The recovery of nickel and iron in untreated concentrate (just grinding and magnetic separation) are 37% and 47.12%. In order to further improve the recovery of nickel and iron, the different factors were investigated.

3.2.1. Effect of Reduction Roasting Temperature

Under the condition of a roasting duration of 60 min, reductant dosage and additive dosage of 10%, and a grinding time of 12 min, the effects of reduction roasting temperature on nickel, iron grade, and recovery in the experimental process were investigated. It can be seen from Figure 4 that the grade and recovery of Ni increase with the increase of the roasting temperature. At 900 °C, the Ni grade and recovery are only 1.79% and 84.59%. When the temperature is raised from 1100 °C to 1250 °C, the Ni grade is increased from 3.04% to 4.01%, and the recovery is increased from 94.67% to 96.75%. The Ni grade increase was obvious. When the temperature is further increased to 1300 °C, the Ni grade reaches 4.39%, and the recovery does not change much. It can be seen that increasing the roasting temperature is beneficial to improve the grade of Ni. However, the roasted samples were found to be partially melted at 1300 °C. The phenomenon is not conducive to reduction roasting operation and increase of the grinding energy consumption. In this experiment, the nickel was enriched by the non-melting reduction as the main research purpose. Therefore, 1250 °C was selected as the subsequent experimental temperature.

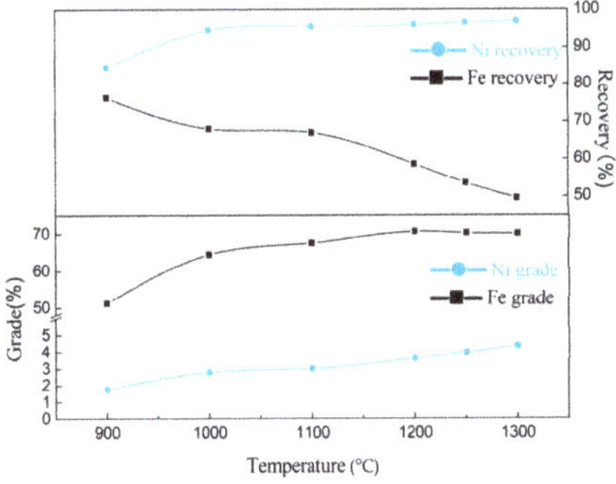

Figure 4. Effect of roasting temperature on the grades and recoveries of Ni and Fe.

In order to further clarify the influence of temperature, the XRD analysis was performed on samples at different roasting temperatures. The results are shown in Figure 5. It can be seen from the XRD pattern that the main phase of the roasting product is forsterite. The existence of the phase is mainly attributed to the following: (1) forsterite is formed by recrystallization of silicate in minerals [20]; (2) serpentine is transformed into forsterite during roasting (Equations (12) and (13)) [21]. No matter how the forsterite is formed, Ni and Fe are released during the phase transformation, which is beneficial to reduction. The intensity of the Fe–Ni diffraction peak in the roasting product was weak at 950 °C, and the FeO diffraction peak was detected. This indicates that Fe and Ni were not fully reduced at the temperature. When the temperature rises to 1050 °C, the intensity of the Fe–Ni diffraction peak obviously increases. Moreover, when the temperature rises to 1250 °C, the Fe–Ni diffraction peak further enhances. The results indicate that the increase of temperature is beneficial to the reduction of Fe and Ni.

$$Mg_3[(Si,Fe)_2O_5](OH)_4 \rightarrow (Mg,Fe)O + (Mg,Fe)Si_2O_5 + H_2O, \tag{12}$$

$$(Mg,Al)_3[(Si,Fe)_2O_5](OH)_4 \rightarrow (Mg,Al,Fe)O + (Mg,Al,Fe)Si_2O_5 + H_2O. \tag{13}$$

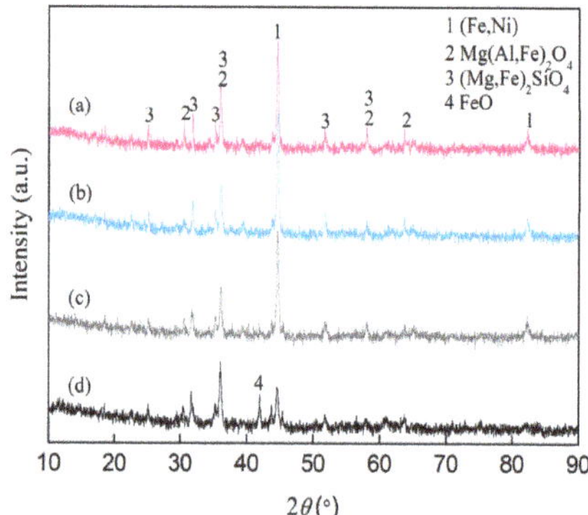

Figure 5. XRD patterns of laterite ore after chloridization and reduction roasting at various temperatures: (a) 1250 °C; (b) 1150 °C; (c) 1050 °C; (d) 950 °C.

In conclusion, the roasting can transform the phase of the ore and the transformation is beneficial to the reduction of nickel and iron. Secondly, the high temperature provides the necessary thermodynamic conditions for the chlorination and reduction reactions.

3.2.2. Effect of Roasting Duration

The experiment was conducted at a roasting temperature of 1250 °C, reductant dosage and additive dosage of 10%, and a grinding time of 12 min. The results are shown in Figure 6. When the roasting duration was 10 min, the Ni grade and recovery were 3.28% and 86.50%. When the roasting duration was extended to 30 min, the Ni recovery reached 99.52% and the Ni grade was increased to 4.40%. When the roasting duration was extended to 80 min, the Ni grade reached 4.96%. In addition, iron recovery was at a low level. Kinetic studies have shown that the reduction of iron is slow and difficult compared to the reduction of nickel, resulting in a higher partial pressure of $FeCl_2$ in the system to inhibit the chlorination of iron oxides [22,23]. In conclusion, the appropriate extension of the roasting duration is beneficial to the recovery of nickel. Considering the nickel grade and recovery, a roasting duration of 80 min is appropriate.

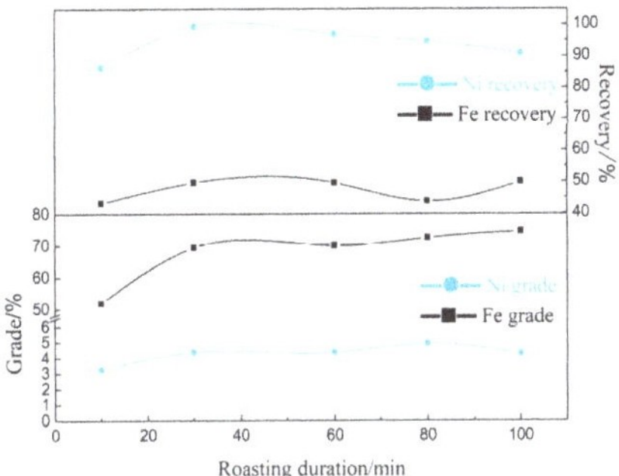

Figure 6. Effect of roasting duration on the grades and recoveries of Ni and Fe.

3.2.3. Effect of Reductant Dosage

Figure 7 shows the effect of the reductant dosage on Ni, Fe grade, and recovery. As can be seen from Figure 7, when the reductant dosage is increased from 2% to 10%, the Ni grade and recovery are increased from 1.61% and 71.67% to 4.96% and 94.81%. Subsequently, with the increase in the amount of the reductant dosage, the Ni grade and recovery showed a downward trend. The added reductant dosage increased the reducing atmosphere. The strong reducing atmosphere is not conducive to the reduction of nickel [24]. In contrast, for the reduction of iron, more reductant dosages are advantageous. Therefore, as the amount of the reductant dosage increased, the Fe grade showed an upward trend. For the sake of comprehensive consideration, the optimum amount of the reductant dosage is determined to be 10%.

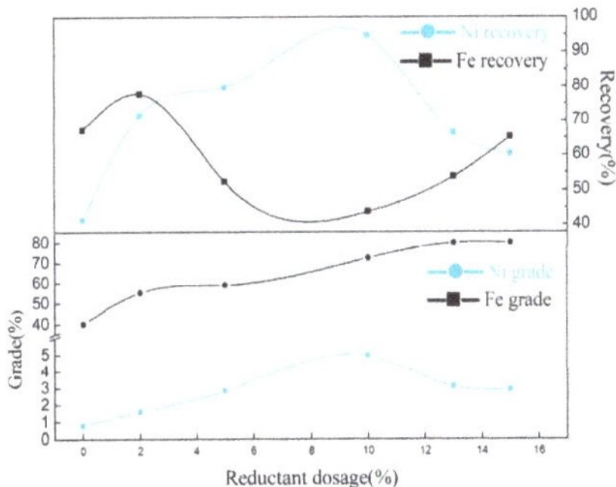

Figure 7. Effect of reductant dosage on the grades and recoveries of Ni and Fe.

3.2.4. Effect of Additive Dosage

In order to study the effect of different additive dosages on the grade and recovery of Ni and Fe, the experiment was carried out under the conditions of a roasting temperature of 1250 °C, a roasting duration of 80 min, a reductant dosage of 10%, and a grinding time of 12 min. The results are shown in Figure 8. In the condition without additive, the Ni grade and recovery were low, only 2.13% and 75.40%. The Fe grade and recovery were 51.12% and 57.13%. With the addition of additives, the Ni grade and recovery increased rapidly. The Ni grade and recovery at 5% of the additive reached 8.15% and 97.76%. This indicates that the addition of appropriate additives promotes the reduction of Ni. Since this study mainly considered the improvement of nickel grade and recovery, the optimum additive dosage was determined to be 5%.

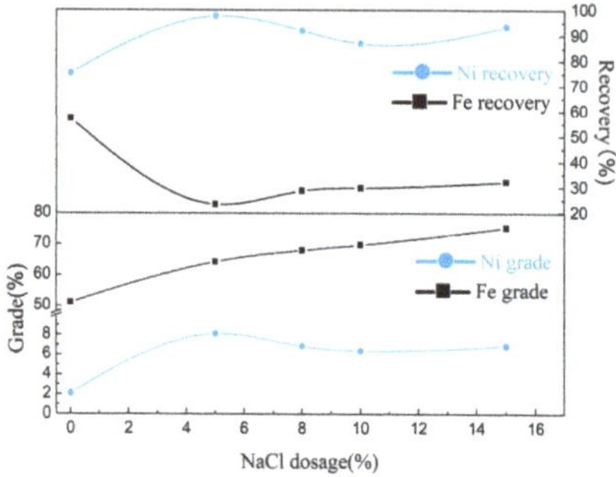

Figure 8. Effect of additive dosage on the grades and recoveries of Ni and Fe.

In order to better compare the effect of the additive on the ore, the roasting product was subjected to XRD analysis. Experimental conditions: roasting temperature 1250 °C, roasting duration 80 min, reductant dosage 10%. The XRD results of roasting products are shown in Figure 9a, (1) no additive and (2) additives with a dosage of 5%. The XRD results showed that there was no significant difference in the phase composition of the two groups, and the forsterite was the main phase. The intensity of the Fe–Ni diffraction peak of (1) is remarkably enhanced as compared to (2). This indicates that the addition of the additive can promote the reduction of Ni and Fe. At the same time, the intensity of the forsterite diffraction peak of (1) is enhanced, indicating the additive promotes the formation of forsterite. The formation of forsterite facilitates the selective enrichment of nickel. The Fe element exists in forsterite and spinel, which is discharged into the tailings during the magnetic separation process, thereby increasing the Ni grade in the concentrate. The XRD pattern of the magnetic separation tailings of the groups (1) (Figure 9b) shows that the main phases in the magnetic separation tailings are forsterite and spinel.

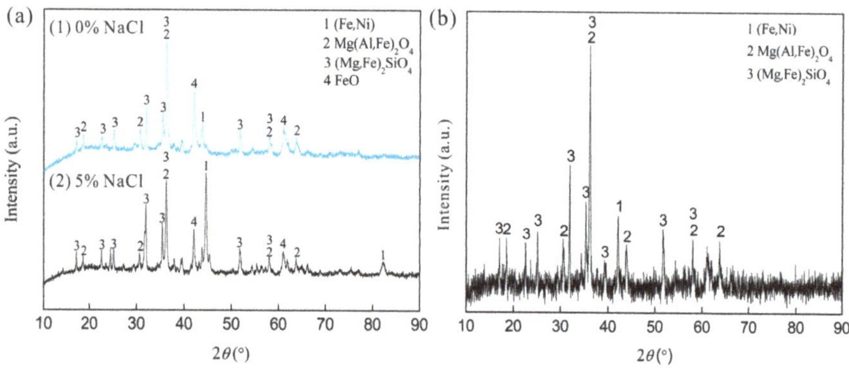

Figure 9. XRD patterns of: (**a**) roasted product; (**b**) tailing.

The roasting product was subjected to scanning electron microscopy (SEM) (Hitachi, Tokyo, Japan) analysis, and the results are shown in Figure 10: (a) no additive, (b) additives dosage 5%. The bright white phase is ferronickel particles. The light gray phase is the spinel and the dark gray phase is the forsterite. The ferronickel particles in Figure 10a are small and dispersed. The size of the ferronickel particles in Figure 10b is remarkably improved, and the ferronickel particles distributed in the roasting product are large. Related literature reports [25] that the chloride is vaporized at high temperature and adsorbed on the carbon surface to be reduced to metal, and then other chlorides are adsorbed on the metal surface to complete the process of reduction and growth. The spinel phase of Figure 10a is significantly reduced and the forsterite phase is significantly increased, indicating that the additive promotes the formation of forsterite. This result is consistent with the XRD analysis of Figure 9a. In addition, Figure 10b shows a sharp crack in the forsterite phase, which is beneficial to the diffusion of reducing gas and metal chloride, and promotes the reduction of Ni and Fe.

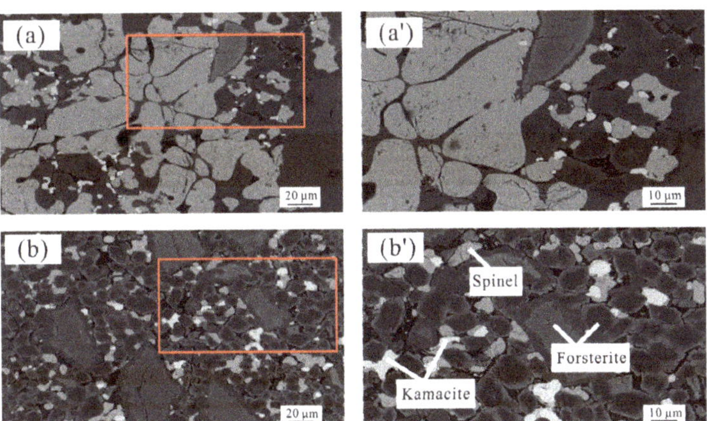

Figure 10. SEM images of roasted ore under two conditions: (**a**) and (**a'**) without NaCl; (**b**) and (**b'**) 5 wt pct NaCl.

3.2.5. Effect of Grinding Time

In order to improve the separation and recovery effect of ferronickel in the magnetic separation process, experimental research on different grinding times was conducted. The experimental conditions were as follows: the roasting temperature was 1250 °C, the roasting duration was 80 min, the reducing agent was 10%, and the additive was 5%. The results are shown in Figure 11. The Ni grade and recovery were 5.78% and 93.89% at 2 min of grinding time. At this time, the roasted product size after grinding is large. The average size of roasted products is D_{90} = 45.25 μm. The results showed that the ferronickel particles were not well separated from the roasting product. With the prolonged grinding time, the Ni grade increased significantly and reached 8.15% at 12 min, at which time the nickel recovery was 97.76%. The average size of roasted products is D_{90} 19.43 μm. The results show that prolonging the grinding time is beneficial to the separation of ferronickel particles from roasted products in the magnetic separation process. However, the prolongation of the grinding time continues, and the size of the roasted product after grinding is further reduced. At this point, the Ni grade and recovery decrease, and the average size of roasted products is D_{90} = 15.11 μm. The reason is that the particle size of some ferronickel particles decreases with the extension of grinding. It is difficult to collect too small ferronickel particles during magnetic separation.

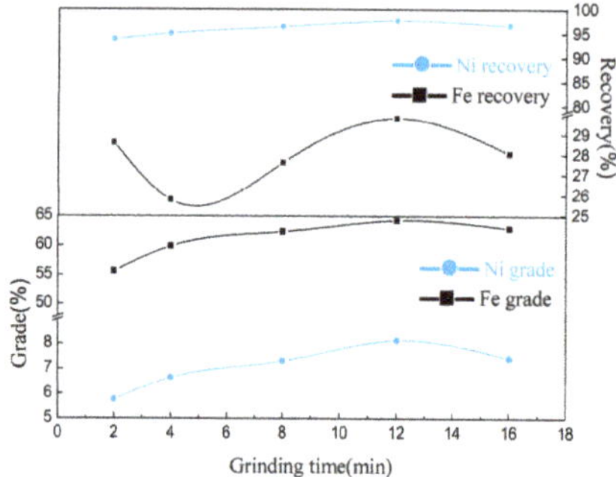

Figure 11. Effect of grinding time on the grades and recoveries of Ni and Fe.

4. Conclusions

The higher nickel grade and recovery were obtained by adding sodium chloride. As a cheap and easily available additive, sodium chloride has obvious advantages compared to other additives. The phase transformation in the roasting process was analyzed. The effect of roasting function on the recovery of nickel and iron was explained.

(1) The addition of sodium chloride as an additive during the reduction process can significantly improve the Ni and Fe grades and recovery in the concentrate. Under the conditions of a roasting temperature of 1250 °C, roasting duration of 80 min, reductant dosage of 10%, additive dosage of 5%, and a grinding time of 12 min, the grade of concentrate Ni and Fe was increased from 1.13% and 51.12% without additives to 8.15% and 64.28%, and the recovery of Ni was increased from 75.40% to 97.76%;

(2) The addition of additives promotes the transformation of the lizardite phase to the forsterite phase, facilitates the dissociation of nickel from the mineral, and improves the reduction effect of nickel. At the same time, the aggregation and growth behavior of ferronickel particles is improved, and the efficiency of magnetic separation is improved.

Author Contributions: Conceptualization, G.Q., S.Z., and H.W.; formal analysis, G.Q.; writing—original draft, G.Q.; writing—review & editing, G.Q., B.L., and Y.W.

Funding: This research was funded by the National Natural Science Foundation of China (No.U1302274) and the International Cooperation Project of Key Research and Development Plan of Yunnan Province (No.2018IA055).

References

1. Sattar, R.; Ilyas, S.; Bhatti, H.N.; Ghaffar, A. Resource recovery of critically-rare metals by hydrometallurgical recycling of spent lithium ion batteries. *Sep. Purif. Technol.* **2019**, *209*, 725–733. [CrossRef]
2. Mudd, G.M. Global trends and environmental issues in nickel mining: Sulfides versus laterites. *Ore Geol. Rev.* **2010**, *38*, 9–26. [CrossRef]
3. Zappala, L.C.; Balucan, R.D.; Vaughan, J.; Steel, K.M. Analysis of a reactive distillation process to recover tertiary amines and acid for use in a combined nickel extraction-mineral carbonation process. *Environ. Prog. Sustain. Energy* **2019**, *38*. [CrossRef]
4. Mudd, G.M. Nickel sulfide versus laterite: The hard sustainability challenge remains. In Proceedings of the 48th Annual Conference of Metallurgists, Sudbury, ON, Canada, 23–26 August 2009.
5. Zhu, D.Q.; Cui, Y.; Vining, K.; Hapugoda, S.; Douglas, J.; Pan, J.; Zheng, G.L. Upgrading low nickel content laterite ores using selective reduction followed by magnetic separation. *Int. J. Miner. Process.* **2012**, *106*, 1–7. [CrossRef]
6. Rao, M.; Li, G.; Jiang, T.; Luo, J.; Zhang, Y.; Fan, X. Carbothermic reduction of nickeliferous laterite ores for nickel pig iron production in China: A review. *JOM* **2013**, *65*, 1573–1583. [CrossRef]

7.   Whittington, B.I.; Muir, D. Pressure acid leaching of nickel laterites: A review. *Miner. Process. Extr. Metull. Rev.* **2000**, *21*, 527–599. [CrossRef]

8.   Jordens, A.; Cheng, Y.P.; Waters, K.E. A review of the beneficiation of rare earth element bearing minerals. *Miner. Eng.* **2013**, *41*, 97–114. [CrossRef]

9.   Keskinkilic, E. Nickel laterite smelting processes and some examples of recent possible modifications to the conventional route. *Metals* **2019**, *9*, 974. [CrossRef]

10. Valix, M.; Cheung, W.H. Effect of sulfur on the mineral phases of laterite ores at high temperature reduction. *Miner. Eng.* **2002**, *15*, 523–530. [CrossRef]

11. Harris, C.T.; Peacey, J.G.; Pickles, C.A. Selective sulphidation of a nickeliferous lateritic ore. *Miner. Eng.* **2011**, *24*, 651–660. [CrossRef]

12. Xiao, F.; Liu, L.; Fang, L.; Yang, R.; Fu, Y.; Zhao, H. Measurement and analyses of molten nickel-cobalt alloy surface tension. *Rare Met. Mater. Eng.* **2008**, *7*, 255–258. [CrossRef]

13. Shen, S.B.; Hao, X.F.; Yang, G.W. Kinetics of selective removal of iron from chromite by carbochlorination in the presence of sodium chloride. *J. Alloy. Compd.* **2009**, *476*, 653–661. [CrossRef]

14. Li, J.; Li, Y.; Gao, Y.; Zhang, Y.; Chen, Z. Chlorination roasting of laterite using salt chloride. *Int. J. Miner. Process.* **2016**, *148*, 23–31. [CrossRef]

15. Okamoto, K.; Ueda, Y.; Noguchi, F. Extraction of nickel from garnierite ore by the segregation-magnetic separation process. *Mem. Kyushu Inst. Technol. Eng.* **1971**, *1*, 41–61.

16. Ilic, I.; Krstev, B.; Stopic, S.; Cerovic, K. The study of chlorination of nickel oxide by chlorine and calcium chloride in the presence of active additives. *Scand. J. Metall.* **1997**, *26*, 14–19.

17. Ilić, I.; Stopić, S.; Cerović, K.; Kamberović, Ž. Study of chlorination of nickel ferrite by gaseous chlorine and calcium chloride in the presence of active additives. *Scand. J. Metall.* **2000**, *29*, 1–8. [CrossRef]

18. Li, X.H.; Zhang, L.X.; Hu, Q.Y.; Wang, Z.X. Effect of phase transformation on chloridizing segregation of laterite ores. *Chin. J. Nonferr. Met.* **2011**, *7*, 238–243. (in Chinese).

19. Fan, C.; Zhai, X.; Fu, Y.; Chang, Y.; Li, B.; Zhang, T.A. Leaching behavior of metals from chlorinated limonitic nickel laterite. *Int. J. Miner. Process.* **2012**, *110*, 117–120. [CrossRef]

20. Li, B.; Wei, Y.G.; Wang, H. Action mechanism and phase transformation characteristics of garnierite in drying process. *Chin. J. Nonferr. Met.* **2013**, *5*, 1440–1446. (in Chinese).

21. Rhamdhani, M.A.; Hayes, P.C.; Jak, E. Nickel laterite Part 1—Microstructure and phase characterisations during reduction roasting and leaching. *Miner. Process. Extr. Metall.* **2009**, *118*, 129–145. [CrossRef]

22. Metallurgical Laboratory of Central-South Institute of Mining and Metallurgy. *Chloridizing Metallurgy;* Metallurgical Industry Press: Beijing, China, 1978; p. 180. (in Chinese)

23. Kanungo, S.B.; Mishra, S.K. Kinetics of chloridization of nickel oxide with gaseous hydrogen chloride. *Metall. Mater. Trans. B* **1997**, *28*, 371–387. [CrossRef]

24. Pickles, C.A.; Forster, J.; Elliott, R. Thermodynamic analysis of the carbothermic reduction roasting of a nickeliferous limonitic laterite ore. *Miner. Eng.* **2014**, *65*, 33–40. [CrossRef]

25. Iwasaki, I. A thermodynamic interpretation of the segregation process for copper and nickel ores. *Miner. Sci. Eng.* **1972**, *4*, 14–23.

 *metals*

Article

The Effect of Titanium Carbonitride on the Viscosity of High-Titanium-Type Blast Furnace Slag

Hongen Xie [1,2], Wenzhou Yu [1,3,*], Zhixiong You [1,3], Xuewei Lv [1,3] and Chenguang Bai [3]

[1] College of Materials Science and Engineering, Chongqing University, Chongqing 400045, China; pzhxiehongen@163.com (H.X.); youzx@cqu.edu.cn (Z.Y.); lvxuewei@163.com (X.L.)

[2] State Key Laboratory of Vanadium and Titanium Resources Comprehensive Utilization, Pangang Group Research Institute Co., Ltd., Panzhihua 617000, China

[3] Chongqing Key Laboratory of Vanadium-Titanium Metallurgy and Advanced Materials, Chongqing University, Chongqing 400044, China; bguang@cqu.edu.cn

*   Correspondence: yuwenzhoucd@cqu.edu.cn; Tel.: +86-23-6511-2631

Received: 8 March 2019; Accepted: 27 March 2019; Published: 30 March 2019

Abstract: In this paper, the effect of titanium carbonitride (Ti(C,N)) on the viscosity of high-titanium-type blast furnace slags was investigated. The different Ti(C,N) contents were achieved by adjusting the reduction degree of TiO_2 to reflect the real characteristics of the high-titanium slag. The results show that the viscosity of the slag increased with the increasing Ti(C,N) content and decreased with the rising temperature. A deviation between the measured and the fitted viscosity appeared as the content of the Ti(C,N) was beyond 4 wt%. Furthermore, the apparent viscous flow activation energy of the slag ranged from 106.13 kJ/mol to 235.46 kJ/mol by varying the Ti(C,N) contents from 0 wt% to 4.97 wt%, which was evidently different from the results of previous studies. The optical microscope and energy dispersive X-ray spectroscopy (EDS) analysis show that numerous bubble cavities were embedded in the slags and the Ti(C,N) particles agglomerated in the solidified samples. This phenomenon further indicates that the high-titanium slag is a polyphase dispersion system, which consists of liquid slag, solid Ti(C,N) particles and bubbles.

Keywords: blast furnace slag; TiO_2; titanium carbonitride; viscosity

---

1. Introduction

Viscosity of TiO_2-containing slags is an important factor influencing the processes of ironmaking, steelmaking and Ti-recycling industries [1–3]. The content of titania (TiO_2) in high-titanium-type blast furnace slag is always more than 20 wt%. Therefore, it is inevitable that part of the TiO_2 in this slag was reduced to titanium carbonitride (Ti(C,N)) particles by coke during blast furnace production. The Ti(C,N) particle is harmful for the fluidity of the slag because it has a high melting point and can significantly increase the viscosity of the slag. According to Einstein's theory [4] about the relation between the volume fraction of solid particles and the viscosity of extremely dilute solutions, Roscoe presented a type of equation (Einstein–Roscoe type equation) to describe the viscosity of liquids containing high concentrations of solid suspensions [5]:

$$\eta = \eta_0(1 - af)^{-n} \tag{1}$$

where η and η_0 are the viscosity of solid-containing and solid-free liquid, respectively; f is the volume fraction of solid particles in the liquid; a and n are constants with regard to the volume fraction and geometrical shape of solid particles in liquid, and are 1.35 and 2.5 for spherical particles with a uniform size, respectively. This equation indicates that the viscosity of the melt should be related to the volume fraction and geometrical shape of solid particles in liquid. To explore this relationship in metallurgical slags, some research has been carried out by adding small amounts of solid particles to the slags. Wright et al. [6] studied the viscosities of CaO-MgO-Al_2O_3-SiO_2 melts containing spinels with different sizes at 1646 K; Liu et al. [7] studied the effect of $Ti(C_{0.3}N_{0.7})$ particles of 1.0 μm on the viscosities of CaO-MgO-Al_2O_3-SiO_2 blast furnace slag and Zhen et al. [8] discussed the effect of TiC particles on the viscosity of CaO-MgO-Al_2O_3-SiO_2-TiO_2 slag. Their results suggested that the viscosity of the solid-containing melt increased with the addition of particles, and the Einstein–Roscoe type equation can well describe the viscosity variation behavior by allowing the parameters a and n to vary. To investigate the flow behavior of high-titanium-type slag, Jiang et al. [9] studied the effect of TiC solid particles on the rheological behavior of blast furnace slags with 20 wt% of total TiO_2 and Yue et al. [10] discussed the rheological behavior of Ti-bearing blast furnace slag with different TiN contents. Both of them pointed out that the slags will convert to non-Newtonian fluids if the volume fraction of the solid particles beyond certain values and the Einstein–Roscoe type equation could be not suitable at that condition.

In the last few decades, a large amount of research [11–19] has attempted to establish an accurate description about the viscosity of the high-titanium-type slag and a lot of fruitful achievements have been obtained. However, the existing empirical and semi-empirical models still cannot describe the viscosity precisely. One of the possible reasons for this may be the improper method for preparing the experimental slag. For example, most of the previous studies prepared the high-titanium-type slag by adding the solid particles directly to the TiO_2-containing slags. However, this could not be enough to reflect the real characteristics of the on-site slag (slag in blast furnace). The morphology and distribution of TiC, TiN, and Ti(C,N) in on-site slag should be different from those directly added to the slag. Additionally, on-site slag is also different from the slag prepared by high purity reagents because there are some gas bubbles in on-site slag, which makes the structure of molten slag more complicated. Up to now, there is still a lack of accurate knowledge of viscosity properties for the high-titanium-type blast furnace slag. In order to control the iron-making process of titanium-vanadium-magnetite more efficiently, a further understanding of flow behaviors in high-titanium-type blast furnace slag should be necessary.

In the present study, to discuss the effect of titanium carbonitride on the viscosity of the high-titanium-type blast furnace slag, on-site slags with different contents of titanium carbonitride (reduced from TiO_2) were prepared. Additionally, the viscosities of these slags were measured to clarify the relationships between slag fluidity and the contents of titanium carbonitride.

2. Materials and Methods

On-site blast furnace slags were used as the raw materials in the experiment. The residual metal iron in the slag was removed by magnetic separation, and then the slag was crushed to less than 0.1 mm. The chemical compositions of the slags were analyzed by X-ray fluorescence spectrometry (XRF, Shimadzu XRF-1800, Kyoto, Japan), as shown in Table 1. Coke was employed as the reductive agent for reducing the TiO_2 to Ti(C,N) in the experiment (compositions are shown in Table 2), which the particle size was controlled between 10 mm and 15 mm.

Table 1. Chemical compositions of on-site slag, mass%.

CaO	SiO_2	MgO	Al_2O_3	TiO_2	TiC	TiN
25.12	24.63	9.50	14.00	22.50	0.49	0.50

Table 2. Chemical compositions of the coke, mass%.

FCad	St	Vadf	Ad	CaO	SiO_2	MgO	Al_2O_3	Fe_2O_3
86.26	0.54	1.23	12.67	0.49	7.01	0.20	3.05	0.95

A muffle furnace (Teenpu CO. LTD., Jiangyin, China) was used to prepare experimental slags with different Ti(C,N) contents. The experimental schematic diagram is shown in Figure 1. The heating element is a U-shaped silicon molybdenum rod, the heating process is controlled by PID (Proportion Integral Differential) program, and the temperature is measured by a B-type thermocouple. About 140 g coke was firstly placed in a graphite crucible (52 mm inner diameter, 160 mm length), and about 210 g slag was placed on the top of the coke. The graphite crucible was then put into a larger corundum crucible to reduce the oxidation degree of the graphite crucible during the experiment. The temperature of the chamber increased to 1773 K at 10 K/min, and then held for 15 min, 30 min, 45 min and 60 min, respectively. After that, the samples were cooled down together with the furnace. The residual cokes in the slags were separated using a hammer and cleaned up by hairbrush. The obtained slags (experimental slags) were crushed to less than 0.1 mm, some of which was used to analyze the chemical composition, and the rest was used to measure the viscosity. The contents of TiC and TiN in the experimental slags were obtained by chemical analysis, and the results are shown in Table 3. The content of Ti(C,N) could not be obtained directly by chemical analysis, thus it was represented by the total contents of TiC and TiN. Number 1 in Table 3 indicates the raw slag without reduction by coke, which the content of Ti(C,N) was 0.99%. The Ti(C,N)-free slag was obtained by roasting the raw slag at 1573 K for 60 min, as represented by 0 in Table 3. The contents of TiC and TiN in this slag were lower than 0.001 mass%, respectively. In order to further confirm the composition of the morphology in the slag, a microstructure analysis was carried out by using the optical microscope (OLYMPUS BX51, Hatagaya, Japan) and scanning electron microscopy with energy dispersive X-ray spectroscopy (SEM-EDS) (FEI, Hillsboro, OR, USA), and the results are shown in Figure 2 and Table 4.

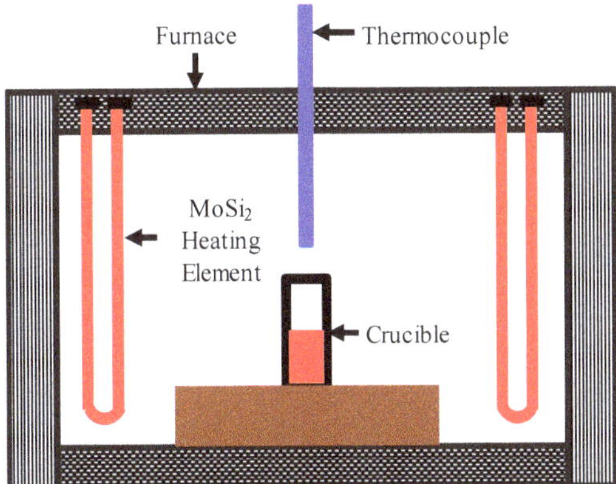

Figure 1. Schematic diagram of the muffle furnace.

Table 3. The contents of TiC and TiN in experimental slags, mass%.

Samples	Time/min	TiC	TiN	Ti(C,N)
0	-	<0.001	<0.001	0
1	-	0.49	0.50	0.99
2	15	1.08	1.09	2.17
3	30	1.79	1.55	3.34
4	45	2.01	2.04	4.05
5	60	2.16	2.81	4.97

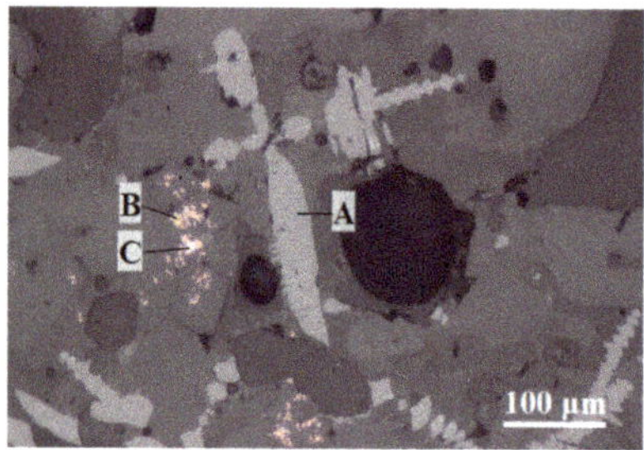

Figure 2. Microstructure of the slag sample (A is Perovskite, B is Tianium carbonitride, C is metal iron.).

Table 4. Energy dispersive X-ray spectroscopy (EDS) analysis of the points in Figure 2, mass%.

Points	C	N	O	Ca	Ti	Fe
Point A	-	-	34.75	29.33	35.92	-
Point B	10.77	5.94	-	-	83.29	-
Point C	-	-	-	-	-	100.0

The EDS (EDAX, Mahwah, NJ, USA) analysis (Table 4) shows that lots of brick-red Ti(C,N) particles were produced in the slag. Additionally, the white iron phase and the grey/white perovskite phase were also observed in the slag.

A schematic diagram of the viscosity measurement apparatus is shown in Figure 3, which includes a Brookfield digital viscometer, a heating system, an automatic lifting system, a temperature control system, and a gas supply system. The heating element is a U-shaped $MoSi_2$ rod, which has a maximum working temperature of 1923 K. The inner diameter of the high purity alundum tube of the shaft furnace is 55 mm, and the height of the constant temperature zone is about 60 mm. The temperature is controlled by PID program, and temperature is measured by a B-type thermocouple with Pt-6 wt pct Rh/Pt-30 wt pct Rh. Argon gas with a purity of 99.99% was employed as the protection gas. During the experiment, the argon gas flowed in from the bottom of the alundum tube and flowed out from the top, and the gas flow was maintained at 1.5 L/min.

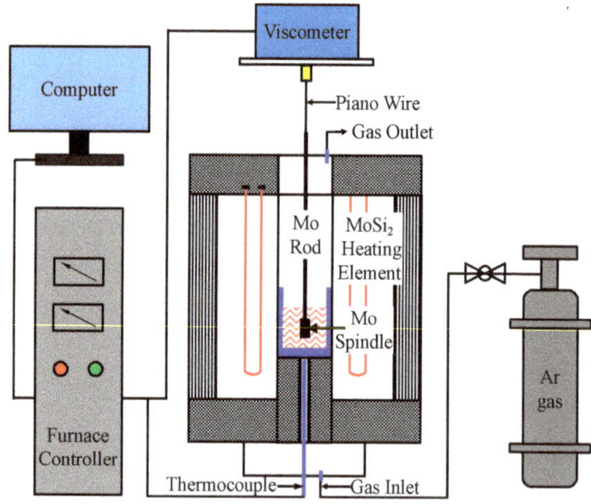

Figure 3. Schematic diagram of the viscosity measurement apparatus.

The viscosity of experimental slag was measured by the rotating-cylinder method. About 170 g experimental slag was put into a high purity graphite crucible (50 mm outer diameter, 40 mm inner diameter, 120 mm length). A metal molybdenum sheet with a thickness of 0.01 mm was put closely inside the graphite crucle to avoid the reaction between slag and graphite. This graphite crucible was put into another larger graphite crucible (62 mm outer diameter, 52 mm inner diameter, 160 mm length) to prevent slag from spilling to damage the alundum tube. The molybdenum spindle (15 mm diameter and 20 mm height) was connected with a molybdenum rod (5 mm diameter and 475 mm length), which linked with the viscometer by a piano wire (carbon spring steel wire). After the molybdenum spindle was slowly immersed into the slag to the predetermined depth, the total height of the liquid slag was about 50 mm. The viscometer was calibrated using three kinds of standard liquids, of which the viscosities were 222.41 mPa·s, 528.20 mPa·s and 1073.3 mPa·s at 293 K respectively. The viscosity was measured when the temperature reached 1773 K. The rotation speed of the spindle was controlled at 12 r/min. The measurement time was 30 min and two viscosity data were obtained per minute, and the average value of measured data was regarded as the viscosity. After the measurement at 1773 K, the temperature was decreased to 1733 K, 1693 K and 1653 K, respectively, and the viscosity at these temperatures was measured by the similar methods.

3. Results and Discussion

3.1. Effect of Ti(C,N) on Viscosity of High-Titanium-Type Blast Furnace Slags

The viscosities of the slags are shown in Table 5. It can be seen that the viscosities of the slags increased as temperature decreased. Simultaneously, with the increase of Ti(C,N) content at the same temperature, the viscosities of the slags increased gradually. This indicates that the viscosity of high-titanium-type blast furnace slags can be influenced not only by temperature, but also by the content of Ti(C,N) particles.

Table 5. Viscosities of the slags, Pa·s.

Samples	1653 K	1693 K	1733 K	1773 K
0	0.194	0.166	0.145	0.123
1	0.289	0.233	0.201	0.170
2	0.570	0.391	0.324	0.272
3	1.052	0.656	0.553	0.452
4	1.794	1.170	0.927	0.783
5	3.178	1.489	1.233	1.017

In the blast furnace ironmaking process of titanium-vanadium-magnetite, it is inevitable that a lot of Ti(C,N) particles are produced. When the Ti(C,N) particles enter the slag, the viscosity of the high-titanium-type blast furnace slag will be influenced significantly. By modifying a and n in Equation (1), or only a and fixed n, the dependence of the measured viscosity on the content of the solid particles could be described by the Einstein–Roscoe type equation. The constant n is related to the geometrical particle shape and can be assumed to be 2.5 for spherical solid particles. The reciprocal value of a represents the maximum amount of solid (f_{max}) that the melt could accommodate before the viscosity becomes "infinite". By fitting the measured values using the Einstein–Roscoe type equation, the values of a can be optimized and the maximum amount of solid (f_{max}) can be obtained, and the results are shown in Table 6. The measured and fitted viscosities of the slags are shown in Figure 4.

Table 6. Values for parameters a, f_{max} and apparent volume fraction at different temperatures.

Parameters	1653 K	1693 K	1733 K	1773 K
n	2.5	2.5	2.5	2.5
a	13.79	12.25	12.02	11.93
f_{max}	0.072	0.082	0.083	0.084

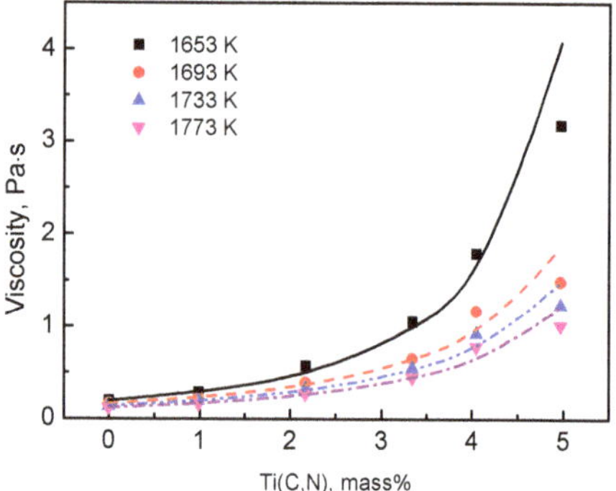

Figure 4. The measured (scattered points) and fitted (lines) viscosities vs. the content of Ti(C,N).

It can be seen from Figure 4 that the higher the content of Ti(C,N) in the slag, the greater the deviation between the fitted viscosity and the measured viscosity. This becomes obvious when the content of Ti(C,N) was over 4 wt%. The main reason for this phenomenon is that the molten slag changed to a non-Newtonian fluid and a shear thinning behavior appeared as the content of the solid particles rose beyond a certain value [9]. It was found from Table 6 that the values of the maximum volume fraction of solids (f_{max}) in this study should be significantly lower than that in the studies of Liu [7] and Zhen [8]. The possible reason for this is that the rotation speed of the spindle in this study (12 r/min) was lower than that used in the other studies (>100 r/min).

However, in addition to the content of Ti(C,N), the influence of bubbles on the viscosity of the on-site slag should not be ignored. It is well known that the smelting of high-titanium-type blast furnace slag can always be accompanied with the formation of bubbles (foam slag). If there are too many bubbles in the molten slag, even if the content of titanium carbonitride is low, the viscosity of the slag will be high [20]. In most cases, the bubbles in the slag were not enough to turn the slag into foam slag. However, there are always more or less bubbles in the slag. The high-titanium-type blast furnace slag should be regarded as a polyphase dispersion system, which consists of liquid slag, solid Ti(C,N) particles, and bubbles. When the foam characteristic value was less than 0.74, the viscosity of the foam can be expressed as [21]:

$$\eta = \eta_0(1 + 4.5\Phi) \tag{2}$$

where η_0 is the viscosity of foam-free liquid and Φ is the foam characteristic value.

Figure 5 shows the microstructure of the solidified slag. It demonstrated that some micro bubble cavities were embedded in the slag, which proves that the bubbles can stably exist in the slag even though the slag is cooled down. This may be another reason that the deviation between the measured viscosity and the fitted viscosity by Einstein–Roscoe type equation.

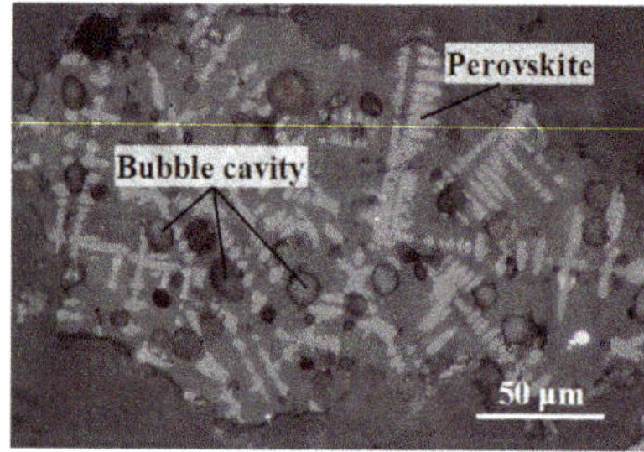

Figure 5. Optical micrograph of the bubble cavity in slag.

3.2. The Effect of Ti(C,N) on the Apparent Viscous Flow Activation Energy E_η

The relation between the viscosity of slag and the temperature was usually expressed by Weymann–Frenkel's equation [22]:

$$\eta = A \cdot T \cdot \exp(E_\eta / (RT)) \tag{3}$$

where A is a proportionality constant, E_η is the apparent activation energy for viscous flow, R is the gas constant and T is the absolute temperature.

According to Equation (3), the curve representing the relationship of $\ln(\eta/T)$–$10^5/T$ is shown in Figure 6. The scattered points are the measured results, and the lines are the fitted results according to Weymann–Frenkel equation. It can be seen that the relationships between $\ln(\eta/T)$ and $10^5/T$ of slags 0 and 1 were almost completely linear, while those for slags 2 to 5 were gradually deviated from the Weymann–Frenkel equation. When the temperature decreased from 1693 K to 1653 K, the differences between the measured values and the fitted values were even greater.

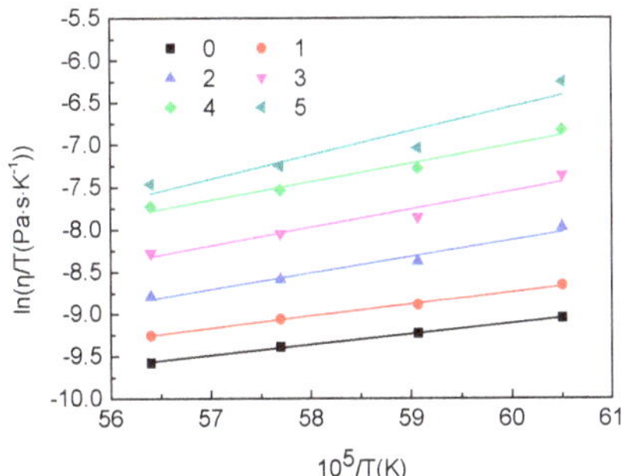

Figure 6. Relationships between $\ln(\eta/T)$ and $10^5/T$ of slags 0 through 5.

The apparent viscous flow activation energies at different contents of titanium carbonitride are shown in Table 7. As the content of titanium carbonitride increases, the apparent viscous flow activation energy increased significantly. This is different from the results of Zhen et al. [8], which showed the apparent viscous flow activation energy experienced no change with increases in the volume fraction of solid particles. It is well known that the apparent activation energy represents the frictional resistance for viscous flow. In Zhen's [8] study, the solid particles were added to the slag, and the interaction between the solid and the slag should be weak. Thus, they suggested that the composition of liquid slag should be the main factor affecting the activation energy of a suspension system. However, in our study, the Ti(C,N) particles in the slag were produced by reducing TiO_2, and thus the interaction between the solid and the slag should be strong. The strong interaction can result in a more complicated structure in the real slag. Therefore, much more energy should be required to overcome the viscous fluid activation energy as the content of Ti(C,N) particles increases.

Table 7. Apparent viscous flow activation energy at different total contents of Ti(C,N), kJ/mol.

Parameters	0	1	2	3	4	5
E_η, kJ/mol	106.13	120.30	161.41	179.73	180.61	235.49
Adjusted R-Square	0.9978	0.9953	0.9605	0.9329	0.9527	0.8440

3.3. The Distribution of Ti(C,N) in Slag

Figures 7–9 show the microstructures of slag samples quenched in different ways (cooling with liquid nitrogen, water cooling, and natural cooling, respectively). The microstructure was analyzed using the optical microscope (OLYMPUS BX51, Tokyo, Japan) and the phase composition was confirmed by SEM-EDS. It can be seen that the perovskite phase, metal iron and Ti(C,N) particles were observed in the solidified samples. It is well known that the perovskite phase has a high melting point and can precipitate preferentially from the melt during the cooling process. When decreasing the cooling rate, the crystal size of the perovskite gradually increased, as shown in Figures 7–9. Additionally, a significant agglomeration of Ti(C,N) particles were seen in the slag no matter which cooling methods were adopted. As compared to the samples of water cooling and cooling with liquid nitrogen, the agglomeration of Ti(C,N) particles was more serious in the natural cooling samples. The solid particles adhered together to form the large agglomeration area. The distribution of titanium carbonitrides in this study were evidently different from those works of Liu, Zhen and Wright et al. [6–8], in which solid particles were evenly distributed in the slag. This implies that the properties of the real slag differs from that of synthetic slag. Additionally, it can be seen in Figures 7–9 that the observed Ti(C,N) particles were always adhered with the embedded metallic iron in the slag. The main reason for this phenomenon is that the TiO_2 in the slag can react with the carbon dissolved in the metallic iron. This also proves that the high-titanium-type blast furnace slag has a complicated structure and the Ti(C,N) particles have a strong interaction with the slag, which should be the main influence on the viscosity of the slag.

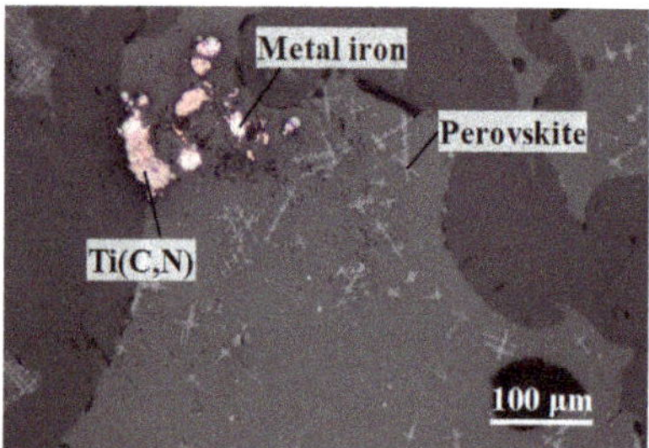

Figure 7. Microstructure of slag sample quenched in liquid nitrogen.

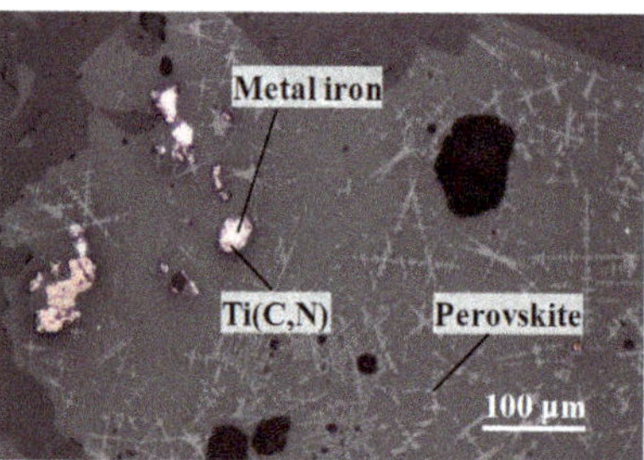

Figure 8. Microstructure of slag sample quenched by water cooling.

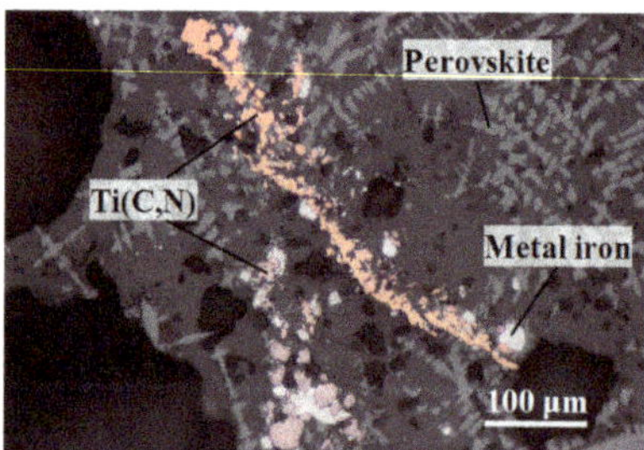

Figure 9. Microstructure of slag sample quenched by natural cooling.

4. Conclusions

This study investigated the viscosity characteristics of high-titanium-type blast furnace slags. Slag samples containing different contents of Ti(C,N) particles were prepared, and the influences of Ti(C,N) on the viscosity of the slag were studied in detail. The main findings can be summarized as follows:

(1) When increasing the content of Ti(C,N), the viscosity of slag samples gradually increased. The measured viscosity and the fitted value by Einstein–Roscoe type equation deviated gradually as the content of Ti(C,N) increased. Additionally, the apparent viscous flow activation energy of the slag also increased with increasing the content of Ti(C,N).

(2) Microbubbles can exist in the slag stably and a serious agglomeration of Ti(C,N) particles was observed in the slag samples. Thus, the high-titanium-type blast furnace slag is actually a polyphase dispersion system, which consists of liquid slag, solid Ti(C,N) particles, and bubbles.

Author Contributions: Conceptualization, W.Y.; Methodology, W.Y.; Software, H.X.; Validation, H.X., W.Y., Z.Y. and C.B.; Formal Analysis, W.Y. and X.L.; Investigation, H.X. and Z.Y.; Resources, W.Y.; Data Curation, W.Y.; Writing—Original Draft Preparation, H.X.; Writing—Review and Editing, W.Y.; Visualization, W.Y.; Supervision, W.Y.; Project Administration, W.Y.; Funding Acquisition, W.Y.

Funding: This research was funded by the National Natural Science Foundation of China (grant number: 51674053, 51704053) and the Fundamental Research Funds for the Key Universities (grant number: 2018CDJDCL0021).

Conflicts of Interest: The authors declare no conflict of interest.

References

1. Hu, K.; Lv, X.W.; Li, S.P.; Lv, W.; Song, B.; Han, K.X. Viscosity of TiO_2–FeO–Ti_2O_3–SiO_2–MgO–CaO–Al_2O_3 for high–titania slag smelting process. *Metall. Mater. Trans. B* **2018**, *49*, 1963–1973. [CrossRef]
2. Wang, Z.J.; Sun, Y.Q.; Sridrar, S.; Zhang, M.; Zhang, Z.T. Investigation on viscosity and nonisothermal crystallization behavior of P–bearing steelmaking slags with varying TiO_2 content. *Metall. Mater. Trans. B* **2017**, *48*, 527–537. [CrossRef]
3. Yang, J.G.; Park, J.H. Distribution behavior of aluminum and titanium between nickel–based alloys and molten slags in the electro slag remelting (ESR) process. *Metall. Mater. Trans. B* **2017**, *48*, 2147–2156. [CrossRef]
4. Einstein, A. *Investigations on the Theory of the Brownian Movement*; Dover Publications, Inc.: New York, NY, USA, 1959.
5. Roscoe, R. The viscosity of suspensions of rigid spheres. *Br. J. Appl. Phys.* **1952**, *3*, 267–269. [CrossRef]
6. Wright, S.; Zhang, L.; Sun, S.; Jahanshahi, S. Viscosity of a CaO–MgO–Al_2O_3–SiO_2 melt containing spinel particles at 1646 K. *Metall. Mater. Trans. B* **2000**, *31*, 97–104. [CrossRef]
7. Liu, Y.X.; Zhang, J.L.; Zhang, G.H.; Jiao, K.X.; Chou, K.C. Influence of $Ti(C_{0.3}N_{0.7})$ on viscosity of blast furnace slags. *Ironmak. Steelmak.* **2017**, *44*, 608–618. [CrossRef]
8. Zhen, Y.L.; Zhang, G.H.; Chou, K.C. Viscosity of CaO–MgO–Al_2O_3–SiO_2–TiO_2 melts containing TiC particles. *Metall. Mater. Trans. B* **2014**, *46*, 155–161. [CrossRef]
9. Jiang, T.; Liao, D.M.; Zhou, M.; Zhang, Q.Y.; Yue, H.R.; Yang, S.T.; Duan, P.N.; Xue, X.X. Rheological behavior and constitutive equations of heterogeneous titanium–bearing molten slag. *Int. J. Miner. Metall. Mater.* **2015**, *22*, 804–810. [CrossRef]
10. Yue, H.R.; He, Z.W.; Jiang, T.; Duan, P.L.; Xue, X.X. Rheological evolution of Ti–bearing slag with different volume fractions of TiN. *Metall. Mater. Trans. B* **2018**, *49*, 2118–2127. [CrossRef]
11. Zhang, L.; Zhang, L.N.; Wang, M.Y.; Lou, T.P.; Sui, Z.T.; Jang, J.S. Effect of perovskite phase precipitation on viscosity of Ti–bearing blast furnace slag under the dynamic oxidation condition. *J. Non-Cryst. Solids* **2006**, *352*, 123–129. [CrossRef]
12. Yue, H.R.; Jiang, T.; Zhang, Q.Y.; Duan, P.N.; Xue, X.X. Electrorheological effect of Ti–bearing blast furnace slag with different TiC contents at 1500 °C. *Int. J. Miner. Metall. Mater.* **2017**, *24*, 768–775. [CrossRef]
13. Zheng, K.; Zhang, Z.T.; Liu, L.L.; Wang, X.D. Investigation of the viscosity and structural properties of CaO–SiO_2–TiO_2 Slags. *Metall. Mater. Trans. B* **2014**, *45*, 1389–1397. [CrossRef]
14. Morizane, Y.; Ozturk, B.; Fruehan, R.J. Thermodynamics of TiO_x in blast furnace-type slags. *Metall. Mater. Trans. B* **1999**, *30*, 29–43. [CrossRef]

15. Liao, J.L.; Li, J.; Wang, X.D.; Zhang, Z.T. Influence of TiO_2 and basicity on viscosity of Ti bearing slag. *Ironmak. Steelmak.* **2012**, *39*, 133–139. [CrossRef]

16. Sohn, I.; Min, D.J. A review of the relationship between viscosity and the structure of calcium-silicate-based slags in ironmaking. *Steel Res. Int.* **2012**, *83*, 611–630. [CrossRef]

17. Gao, Y.H.; Bian, L.T.; Liang, Z.Y. Influence of B_2O_3 and TiO_2 on viscosity of titanium–bearing blast furnace slag. *Steel Res. Int.* **2015**, *86*, 386–390. [CrossRef]

18. Park, H.; Park, J.Y.; Kim, G.H.; Sohn, I. Effect of TiO_2 on the viscosity and slag structure in blast furnace type slags. *Steel Res. Int.* **2012**, *83*, 150–156. [CrossRef]

19. Xi, F.S.; Li, S.Y.; Ma, W.H.; He, Z.D.; Geng, C.; Chen, Z.J.; Wei, K.X.; Lei, Y.; Xie, K.Q. Simple and high-effective purification of metallurgical-grade silicon through Cu-catalyzed Chemical leaching. *JOM* **2018**, *70*, 2041–2047. [CrossRef]

20. Jiang, R.; Fruehan, R.J. Slag foaming in bath smelting. *Metall. Trans. B* **1991**, *22*, 481–489. [CrossRef]

21. Sheng, Z.; Zhang, Z.G.; Kang, W.L. *Colloid and Surface Chemistry*; Chemical Industry Press: Beijing, China, 2012.

22. Urbain, G. Viscosity estimation of slags. *Steel Res.* **1987**, *58*, 111–116. [CrossRef]

Article

Mechanism of Melt Separation in Preparation of Low-Oxygen High Titanium Ferroalloy Prepared by Multistage and Deep Reduction

Chu Cheng [1,2,*], Zhihe Dou [3] and Tingan Zhang [3]

[1] School of materials science and engineering, Henan University of Science and Technology, Luoyang 471003, China

[2] Collaborative Innovation Center of Nonferrous Metals Henan Province, Henan University of Science and Technology, Luoyang 471003, China

[3] School of Metallurgy, Northeastern University, Shenyang, Liaoning 110819, China; douzh@smm.neu.edu.cn (Z.D.); zta2000@163.net (T.Z.)

[*] Correspondence: cheng_chu_love@126.com; Tel.: +86-155-1534-9877

Received: 2 December 2019; Accepted: 14 January 2020; Published: 27 February 2020

Abstract: A novel method to prepare low-oxygen and high-titanium ferroalloy by multistage and deep reduction was proposed in this study. Specifically, the raw materials, high titanium slag and iron concentrate are firstly reduced by insufficient Al powder to obtain high temperature melt. Secondly, CaO and CaF_2 are added into the melt to adjust the basicity of the molten slag. Then, a melt separation under the heat preservation is carried out to intensify the slag-metal separation. Finally, calcium or magnesium is added into the metal melt for a deep reduction. Thereafter, high titanium ferroalloy with an extra-low oxygen content can be obtained. Effects of slag basicity and melt separation time on the slag-metal separation removal were systematically studied. The results indicate that the high titanium ferroalloy, produced by the thermite method, contains a lot of Al_2O_3 inclusions. This leads to a high oxygen and aluminum content in the alloy. With a melt separation with high basicity slag treatment, the Al_2O_3 inclusions can be effectively removed from the alloy melt, and the slag-metal separation efficiency is greatly improved. With the addition of high basicity slag during melt separation, Ti content in the alloy is improved from 51.04% to 68.24%. Furthermore, and the Al and O contents are reduced from 10.38% and 9.36% to 4.24% and 1.56%, respectively. However, suboxides, such as Ti_2O and $Fe_{0.9536}O$, still exist after a melt separation. This indicates that a deep reduction is needed to obtain extra-low oxygen high titanium ferroalloy.

Keywords: multistage and deep reduction; low-oxygen high titanium ferroalloy; inclusions; melt separation; slag-metal separation

---

1. Introduction

High titanium ferroalloy is one of the most important alloys in the melting of special steel and structural steel. It is also an important directive alloy in aviation and space [1]. At present, the major method for the industrial production of high titanium ferroalloy is vacuum remelting process. However, this method has disadvantages with respect to the limited source of titanium scrap and a high production cost [2]. It has been a hotspot issue to directly prepare titanium or titanium alloy from titanium oxides by the electrochemical reduction. Molten salt electrolytic methods, as a short process and direct preparation of titanium alloys, is becoming an issue with an increased concern, while it has not been used in a real production due to the limits of electrode materials and current efficiency [3–8]. Thermite method (also called thermite reduction), using aluminum as a reductant and high-grade ilmenite as raw materials, is another short process of directly producing high titanium ferroalloy. However, this method has some disadvantages of incomplete reduction of titanium oxide due to the weak reducing action of aluminum and incomplete slag-metal separation with a direct casting, which results in a high residual oxygen (up to 12%) and aluminum contents [9,10]. All industrial production lines, using the thermite method to produce high titanium ferroalloy have been shut down. In recent years, researchers in titanium industry have studied the preparation and deoxidation mechanism of high titanium ferroalloy with the thermite reduction method. They all agreed that low-oxygen high-titanium ferroalloy cannot directly be produced from titanium oxides by thermite reduction [11–13]. Chunarev et al. [14] researched technological possibility of manufacturing high-titanium ferroalloy form crude ore. The results show that high ferrotitanium (Ti, 60% to 70%) with oxygen content less than 5% (wt.%) cannot directly be produced by thermite method and the oxygen exists in the forms of Ti_4Fe_2O phase.

Dou et al. [15,16] has investigated the formation mechanism of oxygen in high titanium ferroalloy. The result indicates that the oxygen in high titanium ferroalloy exists in forms of Al_2O_3, suboxides and titanium-oxygen solid solution. He has further researched the effects of physical properties of molten slag on the slag-metal separation and eliminating inclusions in the high titanium ferroalloy prepared by thermite reduction. The results indicate that the appropriate addition of CaF_2 and high basicity of the slag improves the effects of slag-metal separation and inclusion removal [17,18].

Based on our previous researches, a novel methodology for directly preparing low oxygen and high titanium ferroalloy from high titanium slag and iron concentrate by multistage and deep reduction is put forward [19,20]. This method is described as follows: (1) Thermite reduction. High titanium slag and iron concentrate as raw materials are firstly reduced by insufficient Al powder to obtain high temperature melt. (2) Melt separation under heat preservation. CaO and CaF_2 are added into the melt by thermite reduction to adjust the basicity of the molten slag and a melt separation under the heat preservation is carried out to strengthen the slag-metal separation. (3) Deep reduction. Calcium or magnesium is added into the metal melt to carry out a deep reduction and extra-low oxygen high titanium ferroalloy is obtained. In this paper, the effects of slag basicity and melt separation time on the slag-metal separation during the melt separation under heat preservation were systematically studied.

2. Experimental Part

High-titanium slag (86.03% TiO_2, particle size: ≤ 3 mm, Panzhihua Iron & Steel Group Co. China) and iron concentrate (63.07% TFe, particle size: 0.1–0.5 mm, Fortescue Metals Group Ltd, Pilbara, Australia) were used as the raw materials. The chemical composition of the raw materials is shown in Table 1. Aluminum powder (Al, 98.5%, particle diameter: ≤ 3 mm, Jinzhou Metal Co., Ltd., Jinzhou China) was used as the reductant. $KClO_3$ (99.5%, particle diameter: ≤ 3 mm), CaO (99.5%, particle diameter: ≤ 250 μm), magnesium powder (99.5%, particle diameter: 0–0.2 mm) were produced by Sinopharm Chemical Reagent Co., Ltd. City, Shanghai, China.

Table 1. Chemical composition of raw materials, wt.%.

Raw Material	TiO_2	TFe	Al_2O_3	SiO_2	CaO	MgO	Others
High-Titanium Slag	86.03	2.40	3.30	5.60	0.17	0.98	Bal.
Iron Concentrate	-	63.07	2.71	3.40	0.54	0.45	Bal.

The workflow of the experiment is shown in Figure 1. High-titanium slag, iron concentrate, CaO and $KClO_3$ were put into the oven to dry at 110 °C for 24 h, then they were evenly mixed with Al powder in proportion by a barreltype mixer for 40 min. The mass ratio of high-titanium slag, Al powder, iron concentrate, $KClO_3$ and CaO is 1: 0.61: 0.23: 0.35: 0.13, the stoichiometric ratio of Al powder used as the reductant for TiO_2 is 0.9. The reactants were preheated at 150 °C for 2 h, and then were put into a homemade graphite reactor (dia. 350 mm, 480 mm in height). Magnesium powders (about 3 g) were covered with on the top of the reactants and then were ignited to introduce thermite reaction. After a complete thermite reaction, a high temperature melt was obtained and then was teemed into a furnace (SPZ-160A, dia. 150 mm, 200 mm in deep) for a heat preservation. Next, two group of experiments were separately carried out for melt separation, and the experimental process is as follows. Group 1: the high temperature melt was directly carried out a melt separation at 165 °C for 30 min, 60 min, 90 min respectively. Group 2: eliminated 90% of the upper slag and added Al_2O_3, CaO and CaF_2 into the molten slag to adjust the basicity, making its composition that the mass ratio of Al_2O_3, CaO and CaF_2 was 1: 1: 0.05, and then kept the melt at 1650 °C for 30 min, 60 min, 90 min, respectively. After the melt separation, it was cast into a graphite crucible and the alloy ingot was obtained.

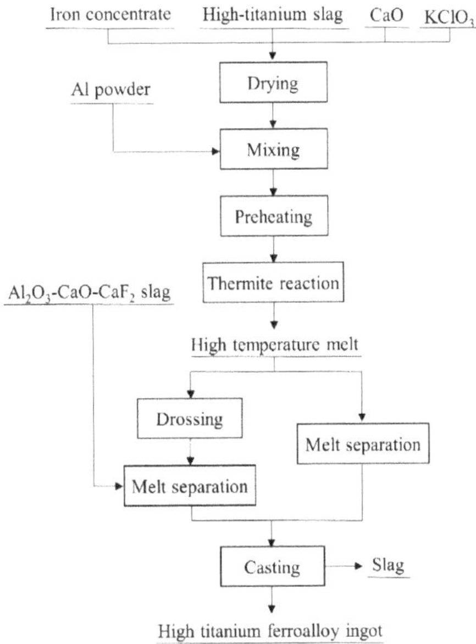

Figure 1. The workflow of the experiment.

The alloy and slag were analyzed by X-ray diffractometry (XRD, Model D8, Bruker, Karlsruhe, Germany; working conditions: Cu Kα1, 40 kV, 40 mA). The composition of high basicity slag after melt separation was analyzed by X-ray fluorescence (XRF, Model S4, Bruker, Karlsruhe, Germany). The oxygen contents of alloy samples were measured by the oxygen and nitrogen hydrogen analyzer (Type G8, Bruker, Karlsruhe, Germany). The compositions of alloys were analyzed by inductively coupled plasma atomic emission spectrometry (ICP, Optima 4300DV, PerkinElmer, Waltham, USA). The microstructures of alloy samples were observed by Scanning Electron Microscope (SEM, SU-8010, Hitachi, Tokyo, Japan) and Energy Dispersive Spectrometer (EDS, Bruker, Karlsruhe, Germany).

3. Results and Discussion

3.1. Characteristics of Slags

Figure 2 shows XRD patterns of slags after melt separation without addition of Al_2O_3-CaO-CaF_2 slag. The diffraction peaks can be indexed to Al_2O_3, TiO, $Ti_{0.913}O_{0.7304}$, $Fe_{3.3}Ti_{9.7}O_3$ and $FeTi_{2.603}O_{0.35}$. The existence of Al_2O_3 indicates that, in thermite method, it is difficult to completely remove Al_2O_3 inclusions from the alloy prepared by thermite reaction with direct casting. The slag-metal separation is strengthened by a melt separation after thermite reaction, but this process will decrease the yield of Ti and Fe. This is due to that TiO, $Ti_{0.913}O_{0.7304}$, $Fe_{3.3}Ti_{9.7}O_3$ and $FeTi_{2.603}O_{0.35}$ will be separated from alloy. The existence of TiO, $Ti_{0.913}O_{0.7304}$, $Fe_{3.3}Ti_{9.7}O_3$ and $FeTi_{2.603}O_{0.35}$ indicates that thermite reaction in this experiment is incomplete. This is mainly attributed to the inadequate aluminum powder (90% stoichiometric ratio of Al powder used as reductant for TiO_2) in ingredients. Therefore, a deep reduction using calcium or magnesium as a reductant is necessary. The titanium sub-oxides and Ti-Fe-O complex oxides in slags indicate that they will be absorbed into the refining slag during a melt separation process, which will decrease the recovery rate of Ti and Fe. In addition, the diffraction peaks intensity of Al_2O_3 are improved with the increase of melt separation time. This means that that Al_2O_3 is effectively removed in the metal melt. Table 2 shows that the content of Al_2O_3 increases with the melt separation time. This result agrees well with the XRF analysis.

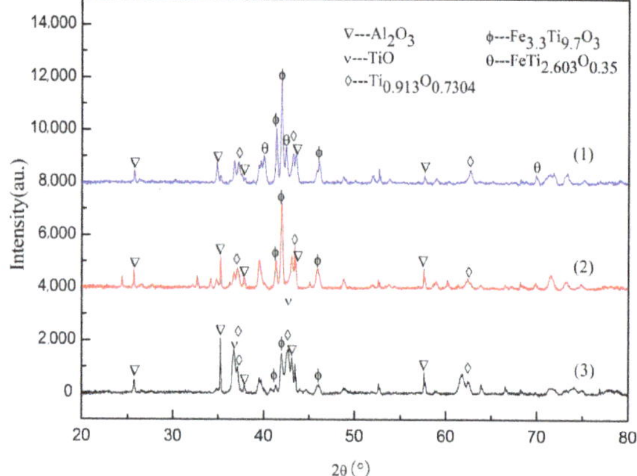

Figure 2. XRD (X-ray diffractometry) patterns of slags after melt separation without addition of Al_2O_3-CaO-CaF_2 slag: **(1)** 30 min; **(2)** 60 min, **(3)** 90 min.

Table 2. XRF (X-ray fluorescence) analysis result of slags after different melt separation time (**1,2,3**: without addition of Al_2O_3-CaO-CaF_2 slag, **4,5,6**: with addition of Al_2O_3-CaO-CaF_2 slag), wt.%.

Number	Separation Time	Al_2O_3	CaO	TiO_2	CaF_2	SiO_2	MgO	Fe_2O_3	Others
1	30 min	12.01	0.64	67.03	-	4.54	0.03	13.39	Bal.
2	60 min	12.56	0.14	67.41	-	3.96	0.01	13.57	Bal.
3	90 min	12.94	0.56	65.97	-	4.04	0.07	14.38	Bal.
4	30 min	49.76	33.31	7.69	4.32	0.94	0.54	0.33	Bal.
5	60 min	52.09	31.06	5.82	5.5	0.81	0.50	0.2	Bal.
6	90 min	53.17	28.15	6.81	5.48	1.24	0.86	0.27	Bal.

Figure 3 shows XRD patterns of slags after melt separation with addition of Al_2O_3-CaO-CaF_2 slag. The diffraction peaks can be indexed to $CaAl_2O_4$ and $CaAl_4O_7$. It can be speculated that high basicity slag reacts with Al_2O_3 inclusions in the alloy melt obtained by thermite reduction and the reaction is as follows [21]:

$$y(CaO)_{slag} + (Al_2O_3)_{inclusion} = (yCaO \cdot Al_2O_3)_{liquid, slag} \tag{1}$$

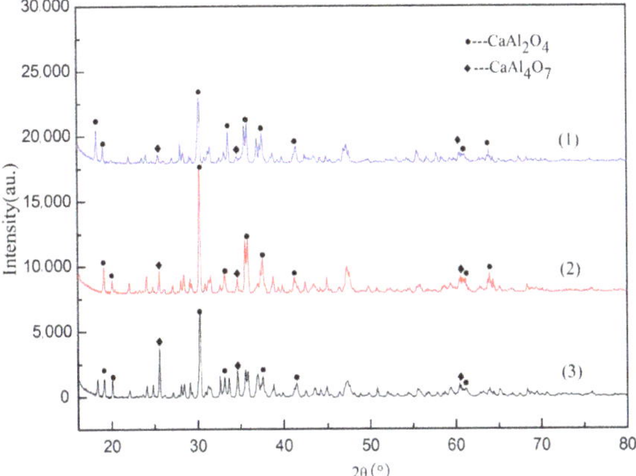

Figure 3. XRD patterns of slags after melt separation with addition of Al_2O_3-CaO-CaF_2 slag at 1650 °C: (**1**) 30 min, (**2**) 60 min, (**3**) 90 min.

Low melting point calcium aluminates are formed during melt separation, and they strengthen the slag-metal separation. Figure 3 shows that the diffraction peaks intensity of $CaAl_4O_7$ increase with the increasing of the melt separation time, while that of $CaAl_2O_4$ increase and then gradually decrease. According to the equilibrium diagram of Al_2O_3-CaO system [22], this result is mainly attributed to that the amount of Al_2O_3 inclusions from alloy melt into the slag increases with melt separation time. Comparing to the XRD patterns of slags without the addition of Al_2O_3-CaO-CaF_2 slag in Figure 2, the diffraction peaks of Ti-Fe compounds oxides and titanium suboxides cannot be observed in Figure 3. The XRF analysis results in Table 2 indicate that the Al_2O_3 content in slag increases with an increased separation time. This indicates that a long melt separation time is beneficial for the slag-metal separation. In addition, the contents of TiO_2 and Fe_2O_3 in the slag with addition of Al_2O_3-CaO-CaF_2 slag are obviously lower than that in the slag without the addition of Al_2O_3-CaO-CaF_2 slag. It can be concluded that the slag-metal separation effect is strengthened, and the migration of alloy components is suppressed by regulating the slag basicity during the melt separation, which is beneficial to increase the recovery rate of Ti and Fe elements.

3.2. SEM Images and EDS Analysis

Figure 4 shows the SEM images of the alloys directly prepared by thermite method and after melt separation under heat preservation.

Figure 4a indicates that the microstructure of the alloy directly prepared by thermite method is loose and contains amorphous Al_2O_3 inclusions with a size about 200 μm, which indicates an incomplete slag-metal separation. Figure 4b–d shows that Al_2O_3 inclusions can also be observed in the alloys after different separation time in the cases without addition of Al_2O_3-CaO-CaF_2 slag. However, the morphology and size of Al_2O_3 inclusions are different with different melt separation time. Comparing to thermite method, Al_2O_3 inclusions in melt separation under heat preservation go through an aggregation-floating upward-recrystallization process, which is beneficial to the Al_2O_3 inclusion removal. The sizes of Al_2O_3 inclusion in the alloy are uneven, ranging from 5–50 μm under a 30 min separation, and the size ranges from 10 to 20 μm after a 60 min separation. However, they are rod-like with a size about 70 μm at 90 min separation. This result indicates that big size Al_2O_3 inclusions can float upwards and are preferentially separated out. However, small size Al_2O_3 inclusions get together, recrystallize and grow up with melt separation time and then are separated out. Compared to the alloys without addition of Al_2O_3-CaO-CaF_2 slag for melt separation in Figure 4b–d, Al_2O_3 inclusions can hardly be observed in the alloys for the cases with addition of Al_2O_3-CaO-CaF_2

slag in Figure 4e–g. In addition, the microstructure of alloys is more uniform and compact, which indicates that the effects of slag-metal separation are greatly enhanced.

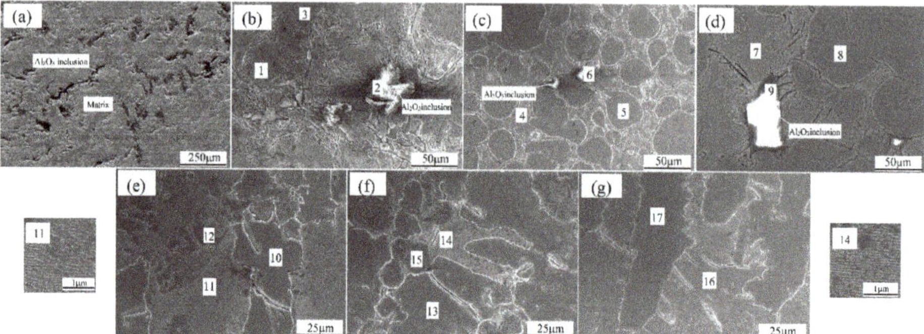

Figure 4. SEM (Scanning Electron Microscope) images of alloys: (**a**) directly prepared by thermite method; (**b–d**) 30 min, 60 min, 90 min without addition of Al_2O_3-CaO-CaF_2 slag; (**e–g**) 30 min, 60 min, 90 min with addition of Al_2O_3-CaO-CaF_2 slag.

According to the microscopic analysis result in Table 3, regions 1, 5 and 8 in Figure 4b–d are supposed to be Ti-O solid solution, and region 3, 4 and 7 are TiFe phase. Si is dissolved in the matrix. Regions 10, 13 and 16 in Figure 4e–g are matrix based on Ti, Fe, Al, Si. Regions 12 and 15 are supposed to be Ti-O solid solutions. Under a high-powered microscope, a "basket microstructure" structure can be observed in the regions 12 and 14 in Figure 4e,f, and they are based on Ti, Fe, Al, Si. However, coarse lath-shaped structures, in the form of titanium silicide, arise in region 17 in Figure 4g. Compared to thermite method, the microstructure of alloys becomes more uniform and denser with a longer separation time.

Table 3. Microscopic analysis of the microstructures of alloys in Figure 4, wt.%.

Region	Ti	Fe	Al	Si	O
1	81.69	2.14	0.78	-	15.39
2	-	-	59.5	-	40.5
3	59.88	20.74	7.94	1.73	9.71
4	53.24	32.52	3.47	2.59	8.18
5	84.5	-	-	-	15.5
6	-	-	58.3	-	41.7
7	55.42	28.49	5.66	2.89	7.54
8	82.15	-	-	-	17.85
9	-	-	61.88	-	38.12
10	42.59	40.25	11.64	3.18	2.34
11	54.88	27.43	5.71	5.38	6.6
12	97.28	-	-	-	2.72
13	43.31	40.72	11.2	3.37	1.4
14	55.61	24.7	7.89	5.43	6.37
15	98.34	-	-	-	1.66
16	44.55	43.12	8.48	2.73	1.12
17	79.42	7.67	4.26	6.65	2

3.3. Phase of Alloys

Figure 5 shows XRD patterns of alloys directly prepared by thermite method with a melt separation. The obvious diffraction peaks of Al_2O_3, $Fe_{3.3}Ti_{9.7}O_3$, $FeTi_{2.6}O_{0.75}$ and TiO can be observed in the alloy directly prepared by thermite method. This means that the metal and slag have an incomplete separation because of the direct casting and a rapid cooling. However, the diffraction peaks of alloys after a melt separation with the addition of Al_2O_3-CaO-CaF$_2$ slag can be indexed to Fe$_{0.9536}$O, TiFe, Ti$_2$O, Al$_3$Ti$_{0.75}$Fe$_{0.25}$, Al$_{86}$Fe$_{14}$ and Al$_{2.1}$Ti$_{2.9}$, but diffraction peaks of Al_2O_3 cannot be observed. The existing state of elements in the alloy is determined by thermite reduction reaction at high temperature, diffusive migration and recombination of constituent element during melt separation [23]. The Ti, which is reduced by residual aluminum in the alloy, preferentially combines with Fe to form Ti-Fe metallic compound and Ti$_{3.3}$Fe$_{9.7}$O$_3$ [24]. However, new intermetallic compounds, such as Al$_{86}$Fe$_{14}$, Al$_3$Ti$_{0.75}$Fe$_{0.25}$, are produced after a melt separation with high basicity Al_2O_3-CaO-CaF$_2$ slag. It can be supposed that the formation of Al$_{86}$Fe$_{14}$, Al$_3$Ti$_{0.75}$Fe$_{0.25}$ is mainly due to the basicity change of slag, diffusive migration and recombination of constituent element during melt separation. As a result, melt separation has a great effect not only on regulating the microstructure of the alloys but also on the removal of Al_2O_3 inclusions. This result agrees well with the microstructure analysis in Figure 4e–g.

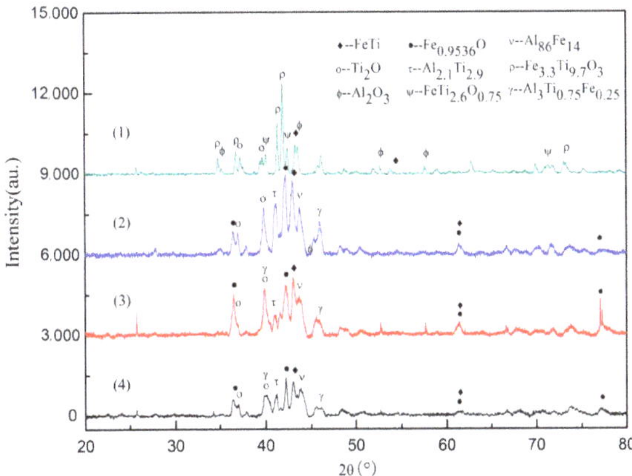

Figure 5. XRD patterns of alloys at 1650 °C: (**1**) directly prepared by thermite method; (**2–4**) 30 min, 60 min, 90 min after melt separation with addition of Al_2O_3-CaO-CaF$_2$ slag.

3.4. Oxygen Content and Chemical Composition Analysis

Figure 6 shows the curves of Al and O contents in the alloy with the melt separation time. The result indicates that both the Al and O contents of the alloys dramatically decrease after a melt separation. The Al and O contents in the alloy directly prepared by thermite method are 10.38% and 9.36%, respectively. The values are reduced to 6.52% and 4.54%, respectively, after a melt separation without the addition of Al_2O_3-CaO-CaF$_2$ slag. However, they are further reduced to 4.24% and 1.56%, respectively, after a melt separation with the addition of Al_2O_3-CaO-CaF$_2$ slag. According to the variation of Al and O before and after melt separation (in accordance with the chemical composition of Al_2O_3), it is supposed that the oxygen mainly exists in the form of Al_2O_3. The removal efficiency of Al_2O_3 during melt separation with Al_2O_3-CaO-CaF$_2$ slag addition is obviously higher than that without Al_2O_3-CaO-CaF$_2$ slag addition. This may be due to the melting point decrease of the slag after the addition of high basicity slag. A low melting point of the slag provides a good condition for a fast metal-slag separation. In addition, the dissolution time of Al_2O_3 inclusion in the high basicity slag is much shorter than the separation time of Al_2O_3 at the slag-metal interface (< 5 s) [25]. At the same time, the dissolution rate of Al_2O_3 particle decreases with the increasing of the Al_2O_3 content in slag, due to that the driving force of Al_2O_3 dissolution decreases. As a result, Al_2O_3 inclusions float upwards and depart from the slag-metal interface. Then, they are absorbed by the melt slag. This should be the main removal procedure of Al_2O_3 inclusion during the melt separation. Therefore, it is important to improve liquidity of melt slag for the removal of Al_2O_3 inclusions. The Al_2O_3 inclusions are effectively removed by strengthening the separation conditions during the melt separation. However, oxygen in the form of suboxides, such as Ti$_2$O and Fe$_{0.9536}$O, still needs to be removed by a deep reduction.

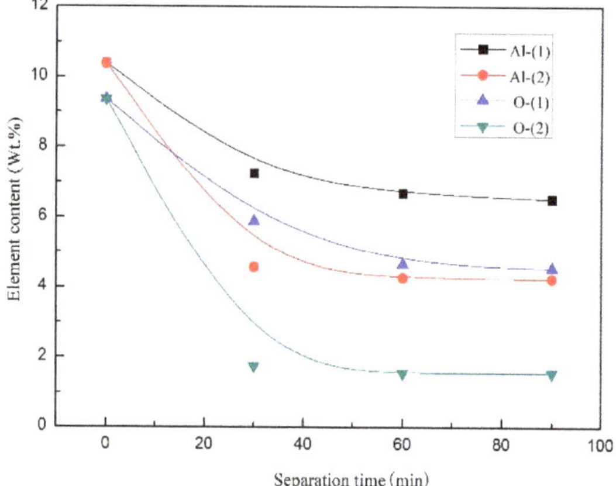

Figure 6. The curves of Al and O contents in alloys with the melt separation time: (**1**) without addition of Al$_2$O$_3$-CaO-CaF$_2$ slag; (**2**) with addition of Al$_2$O$_3$-CaO-CaF$_2$ slag.

Table 4 shows the chemical composition of alloys. The result indicates that the Al and O contents in the alloy directly prepared by thermite method can reach to 10.38% and 9.36%, respectively. After a melt separation without the addition of Al$_2$O$_3$-CaO-CaF$_2$ slag, the Al and O contents dramatically decrease, respectively, from 10.38% and 9.36% to 6.52% and 4.54%, but Ti content evidently increases from 51.04% to 62.12%. After melt separation with addition of Al$_2$O$_3$-CaO-CaF$_2$ slag, the Al and O contents dramatically decrease from 10.38% and 9.36% to 6.52% and 4.54%, respectively. Ti content evidently increases from 51.04% to 68.24%. As a result, melt separation under the heat preservation is the key to produce high titanium ferroalloys with high quality and low impurities. At the same time, the oxygen in the alloy (about 1.50%) after melt separation illustrates that a deep reduction is still needed to further decrease its content, namely a multistage and deep reduction.

Table 4. Chemical composition of high titanium ferroalloys: **1**: directly prepared by thermite method; **2,3,4**: without addition of Al$_2$O$_3$-CaO-CaF$_2$ slag; **5,6,7**: with addition of Al$_2$O$_3$-CaO-CaF$_2$ slag.

Number	Melt Separation Time	Ti	Al	Si	O	Fe
1	-	51.04	10.38	5.65	9.36	Bal.
2	30 min	61.20	7.24	5.26	5.89	Bal.
3	60 min	62.06	6.68	5.47	4.65	Bal.
4	90 min	62.12	6.52	5.36	4.54	Bal.
5	30 min	66.72	4.56	5.08	1.75	Bal.
6	60 min	67.76	4.26	4.64	1.55	Bal.
7	90 min	68.24	4.24	4.86	1.56	Bal.

4. Conclusions

A novel methodology for directly preparing low-oxygen high titanium ferroalloy from high titanium slag and iron concentrate by multistage and deep reduction was proposed, and the effects of slag basicity and melt separation time on the slag-metal separation during the melt separation under heat preservation were systematically studied. The conclusions are as follows:

There are a lot of Al$_2$O$_3$ inclusions in the alloy, which is directly prepared by thermite reduction, and they can be effectively eliminated by a melt separation under a heat preservation. The metal-slag separation efficiency can be effectively improved by increasing the basicity of melt slag and by extending the separation time. After the melt separation with the addition of Al$_2$O$_3$-CaO-CaF$_2$ slag, the Al and O contents dramatically decrease from 10.38% and 9.36% to 4.24% and 1.56%, respectively. The removal mechanism of Al_2O_3 inclusions during the melt separation is that Al_2O_3 inclusions float upwards and depart from the slag-metal interface. After that, they are absorbed by the molten slag.

After a melt separation, the residual oxygen in the form of sub-oxides, such as Ti_2O and $Fe_{0.9536}O$, still needs a deep reduction to be further removed, namely, a deep reduction. The deep reduction process will be the object of subsequent papers.

Author Contributions: C.C. wrote the original manuscript and carried out the tests and data collection. Z.D. supervised experimental work and data analysis. Z.D. and T.Z. revised the manuscript. All authors have read and agreed to the published version of the manuscript.

Funding: This research was funded by"NATIONAL KEY RESEARCH AND DEVELOPMENT PLAN, grant number 2017YFB0305401", "THE NATIONAL NATURAL SCIENCE FOUNDATION OF CHINA, grant number 51422403, 51774078" and "THE FUNDAMENTAL RESEARCH FUNDS FOR THE CENTRAL UNIVERSITIES, grant number N162505002, N172506009, N170908001)".

Acknowledgments: The authors are especially thankful to the professors Yan Liu and Li-ping Niu; they gave us helpful suggestions when the tests were carried out.

Conflicts of Interest: The authors declare no conflict of interest.

References

1. Dou, Z.H.; Fan, S.G.; Cheng, C.; Zhang, T.A.; Shi, G.Y. Research Advance and Application of High Titanniumferrous with Low Oxygen. *Metall. Eng.* **2015**, *2*, 107–121. [CrossRef]
2. Shah, S.J.; Henein, H.; Ivey, D.G. Microstructural Characterization of Ferrotitanium and Ferroniobium. *Mater. Charact.* **2013**, *78*, 96–107. [CrossRef]
3. Qi, C.C.; Hua, X.Y.; Chen, K.H.; Jie, Y.F.; Zhou, Z.R.; Ru, J.J.; Xiong, L.; Gong, K. Preparation of Ferrotitanium Alloy from Ilmenite by Electrochemical Reduction in Chloride Molten Salts. *JOM* **2015**, *68*, 1–7. [CrossRef]
4. Yang, X.; Hu, M.L.; Bai, C.G.; Lv, X.W. Formation Behavior of $CaTiO_3$ during Electrochemical Deoxidation of Ilmenite Concentrate to Prepare Fe-Ti Alloy. *Rare Met.* **2016**, *35*, 275–281.
5. Shi, R.; Bai, C.; Hu, C.; Liu, X.; Du, J. Experimental Investigation on the Formation Mechanism of the TiFe Alloy by the Molten-salt Electrolytic Ilmenite Concentrate. *Int. J. Miner. Metall. Mater.* **2011**, *47*, 99–104. [CrossRef]
6. Du, J.H.; Xi, Z.P.; Li, Q.Y.; Li, Z.X.; Tang, Y. Effect of TiO_2 Cathode Performance on Preparation of Ti by Electro-deoxidation. *Trans. Nonferrous Met. Soc. China* **2007**, *17*, 514–521.
7. Cheng, C.; Zhang, T.A.; Dou, Z.H. Formation Mechanism and Distribution of Al and O in the Ferrotitanium with Different Ti Contents Prepared by Thermite Method. *JOM* **2019**, *71*, 3584–3589. [CrossRef]
8. Liu, X.Y.; Hu, M.L.; Bai, C.G.; Lv, X.W.; Du, J.H. The Deoxidizing Process of the Preparation FeTi Alloy by Electrolysis Ilmenite Concentrate in Molten Salt. *J. Funct. Mater.* **2011**, *42*, 2042–2046.
9. Pourabdoli, M.; Raygan, S.; Abdizadeh, H.; Hanaei, K. A New Process for the Production of Ferrotitanium from Titanium Slag. *Can. Metall. Q.* **2007**, *46*, 17–23. [CrossRef]
10. Dou, Z.H.; Zhang, T.A.; Zhang, H.B.; Zhang, Z.Q.; Niu, L.P.; He, J.C. Basic Research on Preparation of High Titanium Ferroalloy with Low Oxygen Content by Thermit Reduction. *J. Process Eng.* **2010**, *10*, 1119–1125.
11. Tan, S.; Örs, T.; Aydnol, M.K.; Öztürk, T.; Karakaya, İ. Synthesis of FeTi from Mixed Oxide Precursors. *J. Alloys Compd.* **2009**, *475*, 368–372. [CrossRef]
12. Hu, M.; Bai, C.; Liu, X.; Lv, X.I.; Du, J. Deoxidization Mechanism of Preparation FeTi Alloy using Ilmenite Concentrate. *Int. J. Miner. Metall. Mater.* **2011**, *47*, 193–198. [CrossRef]
13. Pazdnikov, I.P.; Dubrovskii, A.Y.; Zelyanskii, A.V. Problem of Optimizing the Production of High-Quality Ferrotitanium. *Titan. Ind.* **2005**, *2*, 21–22.
14. Chumarev, V.M.; Dubrovskii, A.Y.; Pazdnikov, I.P.; Shurygin, Y.Y.; Sel'Menskikh, N.I. Technological Possibilities of Manufacturing High-grade Ferrotitanium from Crude Ore. *Russ. Metall.* **2008**, *6*, 459–463. [CrossRef]
15. Dou, Z.H.; Zhang, T.A.; Zhang, H.B.; Zhang, Z.Q.; Niu, L.P.; He, J.C. Research on Preparation of High Titanium Ferroalloy with Low Oxygen through Vacuum Secondary Reduction Refining. *Rare Met. Mater. Eng.* **2012**, *41*, 420–425.
16. Dou, Z.H.; Zhang, T.A.; Yao, J.M.; Jiang, X.L.; Niu, L.P.; He, J.C. Research on Properties of Al_2O_3-CaO Slag. *J. Process Eng.* **2009**, *9*, 247–249.

17. Shi, G.Y.; Zhang, T.A.; Niu, L.P.; Dou, Z.H. Research on Properties of Low Fluorine CaF_2-CaO-Al_2O_3-MgO-SiO_2 Refining Slag. *J. Process Eng.* **2011**, *11*, 695–697.

18. Dou, Z.H.; Zhang, T.A.; Yao, J.M.; Niu, L.P. Viscosities of Slags in CaO-Al_2O_3-CaF_2-SiO_2 System. *J. Northeast. Univ. Nat. Sci.* **2008**, *29*, 1000–1003.

19. Dou, Z.H.; Zhang, T.A.; Zhang, H.B.; Zhang, Z.Q.; Niu, L.P.; Yao, Y.L.; He, J.C. Preparation of High Titanium Ferrous with Low Oxygen Content by Thermite Reduction-SHS. *J. Cent. South Univ. Technol.* **2012**, *41*, 899–904.

20. Cheng, C.; Dou, Z.H.; Zhang, T.A.; Su, J.M.; Zhang, H.J.; Liu, Y.; Niu, L.P. Oxygen Content of High Ferrotitanium Prepared by Thermite Method with Different Melt Separation Temperatures. *Rare Met.* **2019**, *38*, 892–898. [CrossRef]

21. Han, Z.J.; Liu, L.; Lind, M.; Holappa, L. Mechanism and Kinetics of Transformation of Alumina Inclusions by Calcium Treatment. *Acta Metall. Sin.* **2006**, *19*, 1–8. [CrossRef]

22. Jerebtsov, D.A.; Mikhailov, G.G. Phase Diagram of CaO-Al_2O_3 System. *Ceram. Int.* **2001**, *27*, 25–28. [CrossRef]

23. Zhao, H.M.; Wang, X.H.; Xie, B. Effect of Components on Desulfurization of an Al_2O_3-CaO Base Premolten Slag Containing SrO. *J. Univ. Sci. Technol. Beijing* **2005**, *12*, 225–230.

24. Foadian, F.; Soltanieh, M.; Adeli, M.; Etminanbakhsh, M. A Study on the Formation of Intermetallics during the Heat Treatment of Explosively Welded Al-Ti Multilayers. *Metall. Mater. Trans. A* **2014**, *45*, 1823–1832. [CrossRef]

25. He, S.P.; Chen, G.J.; Guo, Y.T.; Shen, B.Y.; Wang, Q. Morphology Control for Al_2O_3 Inclusion without Ca Treatment in High-aluminum Steel. *Metall. Mater. Trans. B* **2015**, *46*, 585–594. [CrossRef]

Article

Effect of B₂O₃ on the Sintering Process of Vanadium–Titanium Magnet Concentrates and Hematite

Hao Liu [1,2], Ke Zhang [1], Yuelin Qin [1,2,*], Henrik Saxén [3], Weiqiang Liu [3] and Xiaoyan Xiang [1,2,*]

[1] School of Metallurgy and Materials Engineering, Chongqing University of Science and Technology, Chongqing 401331, China; liuhao@cqust.edu.cn (H.L.); m15902303722@163.com (K.Z.)
[2] Value-Added Process and Clean Extraction of Complex Metal Mineral Resources, Chongqing Municipal Key Laboratory of Institutions of Higher Education, Chongqing 401331, China
[3] Process and Systems Engineering Laboratory, Faculty of Science and Engineering, Åbo Akademi University, Biskopsgatan 8, FI-20500 Åbo, Finland; henrik.saxen@abo.fi (H.S.); weiqiang.liu@abo.fi (W.L.)
*   Correspondence: qinyuelin710@163.com (Y.Q.); xxycsu@163.com (X.X.);
    Tel.: +86-1852-392-5702 (Y.Q.); +86-1810-830-9575 (X.X.)

Received: 30 July 2020; Accepted: 3 September 2020; Published: 11 September 2020

Abstract: This work studied the effect of B_2O_3 (analytical reagent) on the parameters of a sintering pot test, as well as the metallurgical properties and microstructure of the sinter samples, to determine the feasibility of applying solid waste containing B_2O_3 in vanadium–titanium sintering. The results show that along with B_2O_3 addition, the mechanical strength of the sinter first increases and then decreases; the maximum strength was found upon the addition of 3.0% of B_2O_3. The low-temperature reduction and pulverization rate of the vanadium–titanium sinter were also improved, while the start and end temperatures of softening showed a decreasing trend. The microstructure of the sinter was found to change from plate structure to particle and point structure, with uniformly distributed small areas. The sintering pots created by B_2O_3 addition had low total porosity but a greater pore diameter than pots created without the reagent.

Keywords: B_2O_3; vanadium–titanium sintering; metallurgical properties; microstructures

---

1. Introduction

Vanadium–titanium magnetite (VTM) is one of the main raw materials for extracting iron and vanadium. At present, the industrial utilization of vanadium–titanium magnetite is still based on the blast furnace-converter route. Sinter containing a high proportion of VTM concentrate exhibits a low drum strength and a high proportion of small-sized particles (<10 mm) [1]. In particular, the fracture toughness of the glass decreases with increasing titanium levels; the low-temperature reduction pulverization rate (RDI$_{-3.15}$) of VTM sinter was 40–60% higher than that of ordinary sintered ore [2,3]. Sinter with these disadvantages has caused many problems in the operation of the blast furnace.

Researchers have tried to improve the metallurgical properties of vanadium–titanium sinter by optimizing the raw materials and processes [4]. Raw material optimization may affect the central parameters and high-temperature performance [5]. The former mainly includes the chemical composition, particle size characteristics, particle morphology and water absorption of the iron ore powders [6]. The latter mainly includes assimilation, liquid phase fluidity, bonding phase strength and melting characteristics [7]. The metallurgical properties of the sinter are determined by the amount of binder phase [8–10]. Gao et al. [11] synthesized different kinds of vanadium–titanium concentrates in laboratory sintering pots to study the effects of the basicity of the raw materials, the moisture content, the particle size of the mixture and the granulation time on the sintering of vanadium–titanium magnetite concentrates. From the perspective of mineralogy, the mass fraction of calcium ferrite in the sinter could be increased by strengthening the granulation [12]. At the same time, the mineral structures were enhanced, and the quality of the sinter was improved [13]. Higuchi et al. investigated the melting formation process via two different experimental techniques, and found that the governing factors for the melt formation of fine ores and coarse ores were different; the quantitative indices of these were estimated individually and re-combined as the sinter strength index to predict the characteristics of the whole ore mixtures [14]. Harvey et al. found that maximum temperature had a strong influence on the porosity of the fired tablets. Over-sintering caused by pore swelling may lead to a decrease in strength. The proposed mechanisms of pore swelling were bubble coalescence and gas generation from hematite decomposition to magnetite [15].

With the environmental problems brought about by wastes containing B_2O_3, such as boron mud, tailings after boron-iron beneficiation, and boric acid waste liquor, the application of these wastes in vanadium–titanium sintering has become an active field of research. Chu et al. [16] showed that indexes such as vertical sintering speed, yield, drum index, sintering cup utilization coefficient, low-temperature reduction and pulverization performance all first increase and then decrease when the mass ratio of boron ore is increased. Hao [17] found that with boric acid and calcium chloride addition, the low-temperature reduction properties and drum strength of the sinter improved, but the high-temperature reduction properties, softening temperature and softening temperature interval did not change. Zhao et al. [18] reported that when a certain amount of boron–iron concentrate was added to the vanadium–titanium sinter, the B_2O_3 could be fully utilized, the iron concentrate grains of sinter grew, the liquid phase increased, and the consolidation strength and metallurgical properties of the sinter were enhanced [19].

From the above studies, it was concluded that B_2O_3-containing materials can improve the sinter properties with vanadium–titanium magnetite as the main raw material. Therefore, the specific role of B_2O_3 has been thoroughly studied. Ren et al. [20,21] found that B_2O_3 mainly deposited in the $2CaO{\cdot}xSiO_2{\cdot}2/3(1-x)B_2O_3$ phase of high-titanium-type vanadium–titanium sinter could effectively inhibit the precipitation of $CaTiO_3$ and $2CaO{\cdot}xSiO_2{\cdot}2/3(1-x)\,B_2O_3$, but promote the precipitation of iron-containing mineral phase. Zhang [22] analyzed the distribution law of the boron element in sinter-pot experiments, and found boron mainly in the glass phase of the sinter.

It is not difficult to see that boron oxide plays an essential role in the sintering process. However, earlier research mainly focused on the effect of materials containing B_2O_3 on the macro-performance of vanadium–titanium sinter. Studies on the influence of B_2O_3 on the sintering process of mixed ore powders using vanadium–titanium fine powder as the main raw material have not been reported. In particular, research on the mechanisms by which B_2O_3 influences the sintering process of the mixed ore powder of vanadium–titanium magnetite and hematite has not been reported so far.

The author earlier studied the effects of B_2O_3 on the assimilation characteristics, softening temperature, fluidity of the liquid phase, compressive strength of the bonding phase and microstructure of the mixed powder of hematite and vanadium–titanium magnetite (H-VTM) [23]. The present work was undertaken to further study the behavior of boron and its effect on sinter structure and quality. Vanadium–titanium magnetite powder and Australian hematite powder were mixed in specific ratios to form an iron ore powder (here called H-VTM) via the production recipe of a company in southwest China. H-VTM and chemical analysis reagent B_2O_3, quick lime and coke powder were used as raw materials for the sinter mix. The effects of B_2O_3 content on the sintering parameters, metallurgical properties and physical transformation of H-VTM powder were studied by using a sintering pot test device. The study was aimed at shedding light on the mechanism by which different B_2O_3 contents affected the sintering and mineralization of the H-VTM mixtures studied.

2. Materials and Methods

2.1. Raw Materials

The iron ore powder used in the experiments was obtained from the raw material site of a steel company in southwest China. For B_2O_3 an analytical reagent ($\geq$98%) was used. The chemical composition of the raw materials is reported in Tables 1 and 2. The raw materials are mixed powder (H-VTM) composed of hematite powder and vanadium–titanium powder in a mass ratio of 2:8. As for the particle size composition of Australian hematite powder and vanadium–titanium refined powder, the grain size <1.0 mm accounted for 42.3% and 57.2%, respectively, while a 1.0–3.0 mm grain size accounted for 12.5% and 8.8%, respectively. The proportions of added B_2O_3 were 0.0%, 1.0%, 2.0%, 3.0%, 4.0% and 5.0%, expressed in wt. %. Each sintered pot test was based on 80 kg of the mixture.

Table 1. Chemical composition of sintered raw materials (wt. %).

Mixture	H-VTM	Coke Powder	Lime	B_2O_3
Ratio, wt. %	85.3	4.2	10.5	0–5

Table 2. Mixture composition.

Mineral	TFe	SiO_2	CaO	Al_2O_3	MgO	TiO_2	FeO	P	S	Loss
VTM	55.78	4.33	0.69	3.86	2.78	9.08	30.50	0.09	0.54	0.5
Hematite	59.76	4.32	0.73	3.16	0.14	0.12	0.80	0.07	0.09	6.40
Lime	-	2.46	84.85	1.53	2.26	-	-	0.02	0.06	1.0
Coke Powder	-	6.0	0.21	4.50	0.8	-	-	0.08	1.1	85

2.2. Methods

2.2.1. Sintering Pot Test Procedure

The leading test equipment used in this work is an experimental sintering set-up, which includes a mixing device, a sintering pot, Frequency modulation (FM) exhaust fan, an ignition combustion fan, a crushing device, a vibration screening and a drum test device (Figure 1). The parameters for the experimental device are given in Table 3.

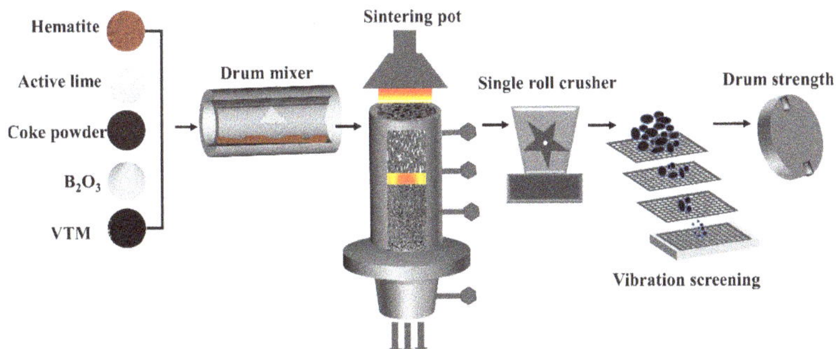

Figure 1. Flow chart of sintering pot test.

Table 3. Parameters of sintering pot.

Sintering Test Parameters	Value	Sintering Test Parameters	Value
Mixture Thickness	650 mm	Sintering Pot Diameter	300 mm
Underpressure (at the beginning)	8 kPa	Underpressure (Constant Level)	16 kPa
Ignition Temperature	1150 °C	Bottom Sinter Thickness	25 mm
Ignition Time	2 min	Granulation Time of first Mixing	2 min
Ignition Gas	Liquefied petroleum gas	Granulation Time of second Mixing	3 min
Pressure Control	FM exhaust fan	Basicity of Mixture	2

2.2.2. Metallurgical Performance Experiment

After undertaking the sintering pot test, the samples were broken, and screened samples were taken according to a certain proportion of particle size distribution to form a comprehensive sample, which were then subjected to low-temperature reduction, pulverization, reduction and soft melt dropping (Figure 2) to study the effect of B_2O_3 on the metallurgical properties of the sinter. These tests are described in the following subsections.

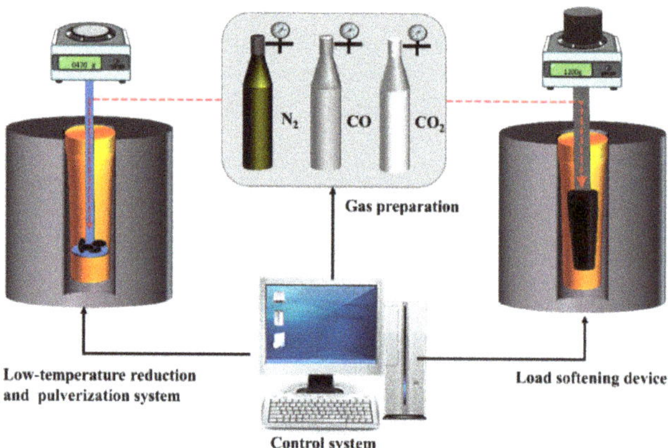

Figure 2. Schematic diagram of the metallurgical performance experimental device.

- Low-temperature reduction pulverization experiment

For sinter samples, this was done according to the national standards of China (GB/T 13242-2017 Iron ores-Low-temperature disintegration test method using cold tumbling after static reduction) [24]. The low-temperature reduction pulverization index is generally expressed as $RDI_{+3.15}$. A sinter sample with a mass of 500 ± 0.5 g and grain size of 10.0–12.5 mm was first placed in a dryer and kept there for 2 h at a temperature of 105 ± 5 °C. After this, the sample was heated at a heating rate of 10 °C/min under a protective atmosphere provided by introducing 5 L/min of N_2 into the reduction tube. When the temperature reached 500 °C, the flow of N_2 was increased to 15 L/min. The temperature was kept constant for 30 min, and during this time the gas distribution device was applied to keep the gas flow stable. After this, to start the reduction reactions, the proportions of reducing gas were 20% (CO), 20% (CO_2) and 60% (N_2). The reduction gas of 15 L/min $\pm$ 0.5 L/min was added instead of inert nitrogen. This reaction environment was maintained at a constant temperature for 1 h. At the end of the reaction, protective N_2 was introduced again, and the sample was cooled naturally at a flow rate of 5 L/min until the temperature was below 100 °C. Finally, the samples were extracted for analysis.

- Reduction index experiment

According to the national standards of China (GB/T 13241-2017 Iron ores-Determination of reducibility) [25], the reducibility measurements of the sinter samples containing B_2O_3 were undertaken in a fixed-bed reactor, under a reducing atmosphere at a high temperature. The reduction degree of the sinter was determined by the weight-loss method. First, the sinter with a grain size of 10.0–12.5 mm was sieved out to obtain a sample of 500 ± 0.5 g. This sample was placed in a blast dryer and kept there 2 h at 105 ± 5 °C. After this, the sample temperature was raised by imposing a heating rate of 10 °C/min under a protective atmosphere (N_2, 5 L/min). When the temperature in the furnace reached 900 °C, the temperature was kept constant for 30 min. Another valve was opened, letting in 30% CO and 70% (N_2) at a flow rate of 15 L/min., simultaneously with turning on the automatic monitoring system for data logging. At the end of the reaction, the reducing gas valve was closed, instead of letting in N_2, and the sample was cooled at a flow rate of 5 L/min until the temperature was below 100 °C. Finally, the sample was extracted for examination.

- Softening and drip test under load

The softening under load was assessed to obtain information about the high-temperature behavior of the sinter. According to the national standards of China (Method for the determination of iron reduction softening dripping performance under load) [26], the test was undertaken by placing a 50 mm layer of sinter (particle size 10–12.5 mm) on 20 g of coke, and adding 15 g of coke on the top. A load of 1.0 kg/cm^2 was applied to the bed and 15 L/min of reducing gas (CO/N_2 = 30/70) was fed through the bed to heat it at a rate of 10 °C/min up to 900 °C. Then the temperature was kept constant for 60 min at 900 °C, followed by further heating (5 °C/min) until the first droplets were collected under the lower coke layer. N_2 was introduced to keep the sample below 400 °C, while reducing gas was introduced at higher temperatures. During heating, the temperature at which the height of the column experienced shrinkage by 10% was noted as the softening start temperature (T_a), the temperature at which 40% shrinkage occurs as the softening end temperature (T_e), and the softening temperature interval was determined as $\Delta T_{ea} = T_e - T_a$. Finally, T_m, the melting temperature, is defined as the temperature at which the first slag/iron droplets are observed under the lower coke layer.

3. Results and Discussion

3.1. Experimental Indicators of H-VTM Sinter of Different B_2O_3 Addition

The results of the tests undertaken in this work revealed that B_2O_3 significantly affected the key properties of the sinter. Figure 3 shows that with the increase in B_2O_3, the drum strength of the sinter increased rapidly from the initial level of 48% (0% B_2O_3) to reach a maximum value of 58.0%, when the B_2O_3 ratio was 3.0%, but decreased with the further addition of B_2O_3 to 46.7% at 5.0% B_2O_3. The vertical sintering speed increased first, then decreased, also showing a maximum at 2.0% B_2O_3. Within the range of 0.0–3.0% B_2O_3 addition, the qualified rate of sintered samples generally improved, but with B_2O_3 > 3.0%, both parameters decreased.

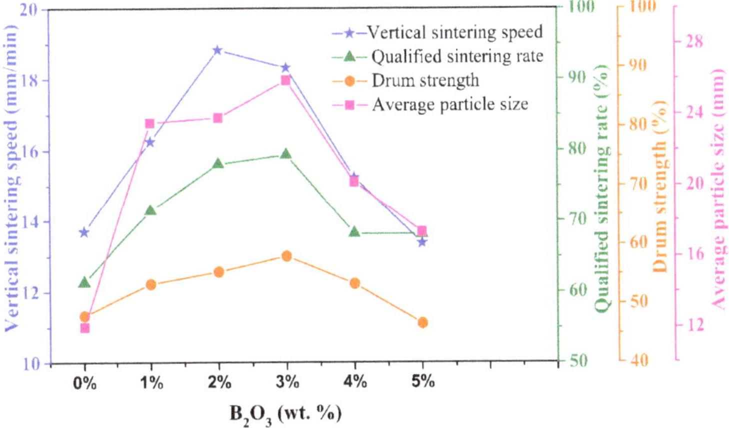

Figure 3. Key sinter properties with different additions of B_2O_3.

The overall average particle size first increased and then decreased, which is illustrated in Figure 4. Figure 5 illustrates that the particle size distribution of the sinter changed significantly with the addition of B_2O_3. The test with 1% addition of B_2O_3 reduced the share of particles <5.0 mm from 39% to 25%, and the share of >40 mm particles increased from 3% to 21%. This in practice means that the content of the bulk sinter increased strongly, and the overall particle size of the sinter also increased. A likely reason for the observed changes is the enhanced adhesive effect caused by the increase in the liquid phase content of the sintering sample along with the increasing levels of B_2O_3. To analyze this from a theoretical perspective, the equilibrium composition was estimated by FactSage software. The results demonstrated that the liquid phase region expands when B_2O_3 is added [23]. In most of the liquid phases, B_2O_3 reacted with some of the oxides in the raw material mix to generate components with low melting points, such as $2CaO \cdot B_2O_3 \cdot SiO_2$, $CaO \cdot B_2O_3 \cdot 2SiO_2$, $CaO \cdot 2B_2O_3$, $CaO \cdot B_2O_3$, $MgO \cdot 2B_2O_3$, $2Al_2O_3 \cdot B_2O_3$, $CaO \cdot B_2O_3 \cdot Al_2O_3$ or $2CaO \cdot B_2O_3 \cdot Al_2O_3$. All these substances have melting points below 1200 °C, and hence can form liquid phases at the temperatures of the experiments [27].

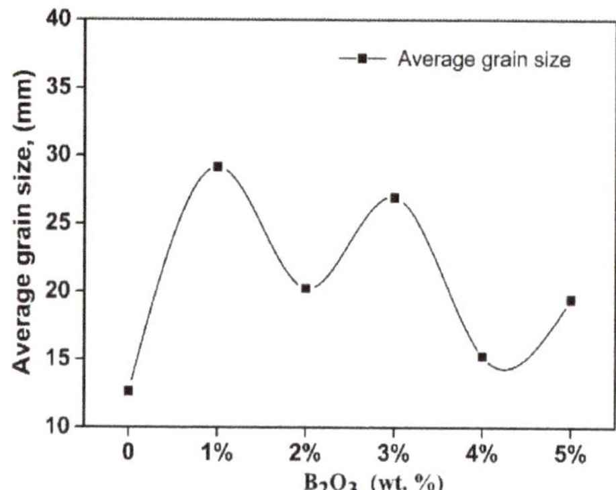

Figure 4. The average grain size of sinter with B_2O_3 addition.

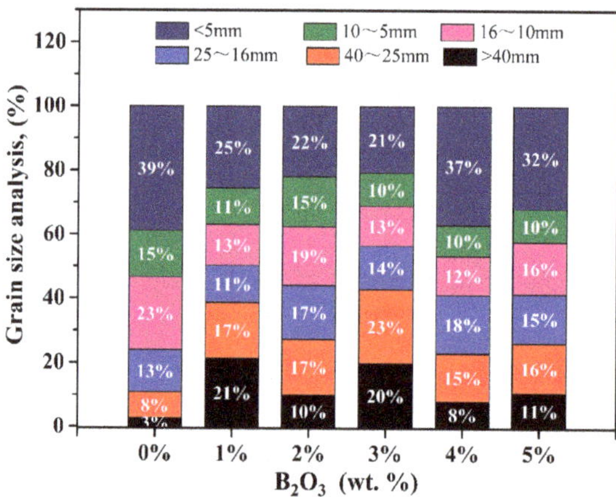

Figure 5. The size distribution of sintered ore in the sintering test for B_2O_3 addition.

When the vanadium–titanium sinter sample is cooled to 725 °C, the internal α'-$2CaO$-SiO_2 is converted to γ-$2CaO \cdot SiO_2$, and its volume increases by approximately 12%. When cooled to 525 °C, β-$2CaO \cdot SiO_2$ changes to γ-$2CaO \cdot SiO_2$. During the transformation of the crystal lattice, rearrangement occurs and the density changes. Hence, the volume increases by approximately 10% during the crystal transformation. Owing to the volume expansion at cooling, the sinter is exerted by a strong force. In severe cases, this internal stress may even pulverize the sinter particles. Thus, a reduced strength may be caused by the existence of various minerals with different thermal expansion coefficients, melting points and crystallization capabilities.

The melting point of B_2O_3 is 450 °C, and its boiling point is 1860 °C. Owing to the low melting point, the addition of B_2O_3 before sintering can promote the generation of liquid crystal nuclei. Furthermore, since the radius of the boron ion is small (0.02 nm), it can diffuse quickly into the crystal lattice of calcium orthosilicate, thereby stabilizing the crystal form of the orthosilicate, reducing the natural pulverization rate and improving the overall strength and particle size of the sample. Along with an increase in B_2O_3, the liquid phase increases. During the cooling phase, the liquid phases shrink and crystallize, and the over-melted sintered liquid phases may also froth because of the reduced surface expansion [28]. This finding applies to higher B_2O_3 contents: contents above 3% lead to large holes and thin-wall structures in the particles, as shown in Figure 6.

Figure 6. H-VTM sintered ore samples with different B_2O_3 addition.

3.2. Changes in the Metallurgical Properties of the Samples for Different B_2O_3 Contents

Before B_2O_3 was added, the low-temperature reduction pulverization index was 73.0%, but an increase of 1.0% in B_2O_3 resulted in an increase of 16.8 points to 89.8%, as shown in Figure 7. However, a further increase in the B_2O_3 content did not significantly improve the index. The addition of B_2O_3 decreased the low-temperature reduction of the smallest (−0.5 mm) fines significantly, from 11.5% to 2.6%, and thus it enhanced the ability of the sinter to resist low-temperature reduction pulverization and significantly reduced the amount of powder at low temperatures. The primary process of sintering was completed in less than 5 min, and the properties and quantity of the primary liquid phase during the sintering of vanadium and titanium were found to affect the formation of the structure of the mineral phase. The viscosity of the liquid phase is a critical factor affecting the reaction rates. The presence of B_2O_3 decreases the viscosity of the liquid phase. The liquid phase shrinks sharply during the cooling process, giving rise to pores. The effect may be indirect on pore volume and the intergranular distance of grains, as the permeability is increased and the coke burning reactions proceed more effectively. The permeability of the combustion zone was improved. This phenomenon was also beneficial to the diffusion of Ca^{2+} to the Fe_2O_3 surface in the liquid phase, allowing for the natural formation of calcium ferrite. The reduction ability and the strength of the calcium ferrite liquid phase were better than those of the silicate liquid phase and perovskite, and improved the overall resistance to pulverization of the vanadium–titanium sinter.

When more than 1.0% of B_2O_3 was added, the reducibility deteriorated clearly, since excessive liquid phases developed and the structure of the sinter became dense. The interior of the sinter melted and consolidated, and the porosity decreased, thus reducing the contact area with the reducing gas, which decreased the reduction rate.

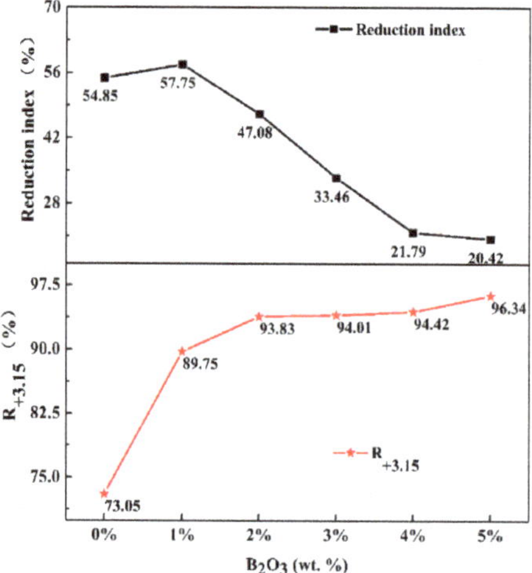

Figure 7. Metallurgical properties of sinter at B_2O_3 addition.

Figure 8 shows that after B_2O_3 addition, the melting start and end temperatures of the sinter sample decreased significantly, the melting temperature interval increased considerably, and the dripping temperature decreased. When no B_2O_3 was added, the melting temperature interval of the sinter sample was 104 °C, which increased by 110 °C with 5% B_2O_3 addition. Thus, the melting interval is strongly affected by the addition of B_2O_3 to the mixture. The reason for this is that as some free B_2O_3 exists in the sinter, its low melting point and excellent dissolution performance make it react with various metal oxides to form low melting-point compounds.

These changes are expected to give rise to an increased thickness of the cohesive zone in the blast furnace, which increases the gas flow resistance. This is not desirable since it may negatively affect the stability of the operation and the CO utilization ratio. Besides, the fact that both the softening start and end temperatures decrease causes an elevation of the vertical level of the cohesive zone, which would cause an increase in the fuel rate of the blast furnace.

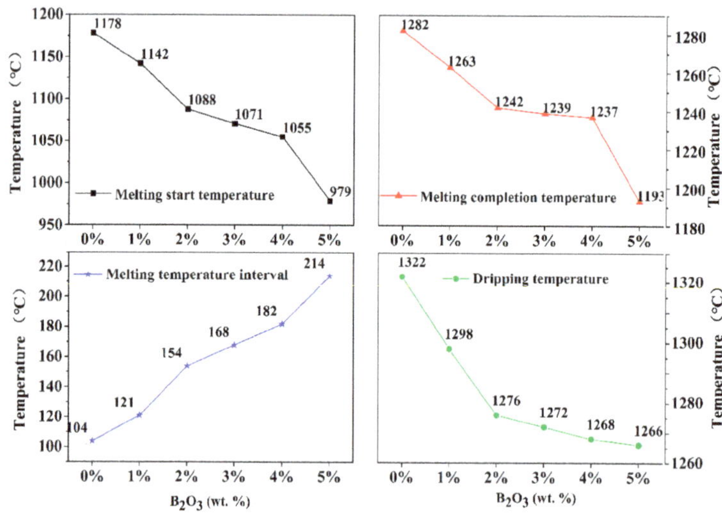

Figure 8. Melting and dripping temperatures of samples with B_2O_3 addition.

3.3. Changes in the Microstructure of the Samples for Different B_2O_3 Contents

The addition of B_2O_3 changes the sinter from a non-uniform ore phase (point, granular or dendritic) to a uniform phase (plate or block). Furthermore, B_2O_3 contributes to the formation of a liquid phase with a low melting point. As such, the amount of liquid phase generated in the crystalline phase of the sinter increases as the amount of liquid phase is increased. The enhanced liquid phase fluidity increases the scope for melting other minerals. At the same time, the presence of enough liquid phase also enhances the internal bonding ability of the sinter. The microstructure of the sinter also becomes tight, thus improving the quality of the sinter, as seen in Figure 9.

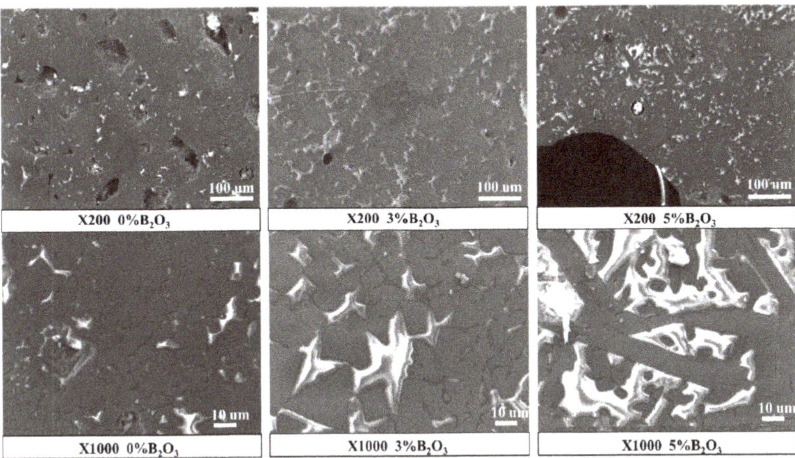

Figure 9. Scanning electron microscope (SEM) images of H-VTM at B_2O_3 addition.

When no B_2O_3 was added, the microstructure of the sinter showed poor homogeneity. Cracks were unevenly distributed and the crystal size was 0.5–1.0 mm. Another crystal phase was magnetite, appearing as small, semi-euhedral crystals with a size in the range 0.5–0.8 mm. Part of the magnetite was bound into a grain structure by dicalcium silicate, calcium, magnesium olivine and a glass phase. The calcium ferrite was plate-shaped and unevenly distributed between hematite and magnetite. A small amount of perovskite appeared and filled the region between the dendritic and magnetite particles. Dicalcium silicate was cemented between the hematite and magnetite as a dendritic phase. With the increase in the B_2O_3 content, the microstructure of the sinter changed significantly, and the mineral structure appeared uniform. The mineralogical structure of samples containing 1.0–3.0% B_2O_3 was mainly granular, partly speckle-granular, and with a skeletal structure. The hematite crystal grain size became small. The crystal grain size was 0.20–0.50 mm, and the overall porosity became small and partly uniform. However, the pore size was large and the number of cracks decreased. Hematite and magnetite were surrounded by large amounts of dicalcium silicate, and the amount of bound phase increased significantly. After adding more than 4.0% of B_2O_3, many pores appeared in the sinter, and the pores were mostly connected. When the amount was further increased, part of the hematite was found to be present as idiotypic crystals, the gaps between the grains were enlarged, and the number of bonding phases increased significantly.

As shown in Figure 10, the main elements of the mineral structure include O, Si, Ca, Ti, Fe and B, and with B_2O_3 addition, B was naturally also detected in the sinter; however, EDS surface scanning could not accurately determine its surface distribution because the atomic weight of boron is very low. The iron phase was mainly magnetite and hematite and filled with perovskite, silicate and ferrite. Mg, Al and O exhibited an even distribution in the mineral phase, and Si and Ca in the bright white area between iron the phases.

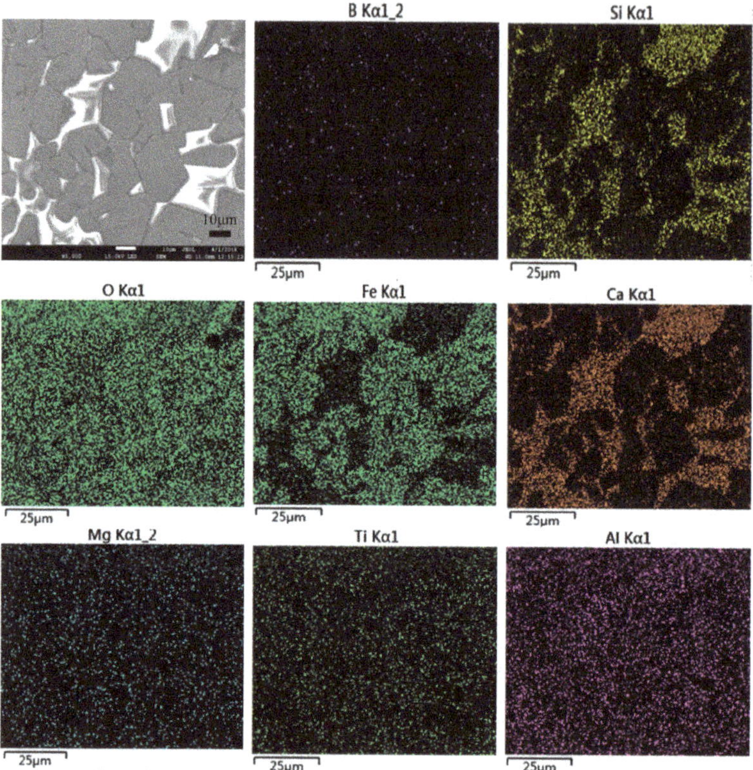

Figure 10. Surface scanning of sample with 5% B_2O_3 added.

At points A and B in Figure 11, the main elements are as shown in Table 4. The Ti content at point A was high and thus was preliminarily judged to be primary titanium magnetite. B represents the grey filled area between the iron oxide phases, and is a common area of dicalcium silicate, calcium ferrite and perovskite with enriched boron. With the gradual increase in the B_2O_3 ratio, the hematite or ilmenite gradually changed from an initial lamellar- and plate-like shape to a granular- and worm-like shape. The structure also changed from non-uniformly distributed large blocks to a uniformly distributed interlaced structure. The distribution of ferrite and titanium iron compounds was very uniform, and the structure gradually changed from a large area of plate structure to a small area of granular and dot structure. Boron was enriched in a common area of dicalcium silicate and composite calcium ferrite. Therefore, B_2O_3 significantly improved the fluidity of the binder phase and increased the total number of cohesive phases.

Table 4. SEM-EDS element content of a sample containing 5.0% of B_2O_3.

Point	Atomic Fraction (%)									
	B	C	O	Mg	Al	Si	Ca	Ti	V	Fe
A	1.17	5.32	26.41	0.96	3.06	0.19	0.23	3.49	0.38	58.80
B	10.19	13.01	37.84	0.97	2.72	8.83	15.67	1.96	0.22	8.59

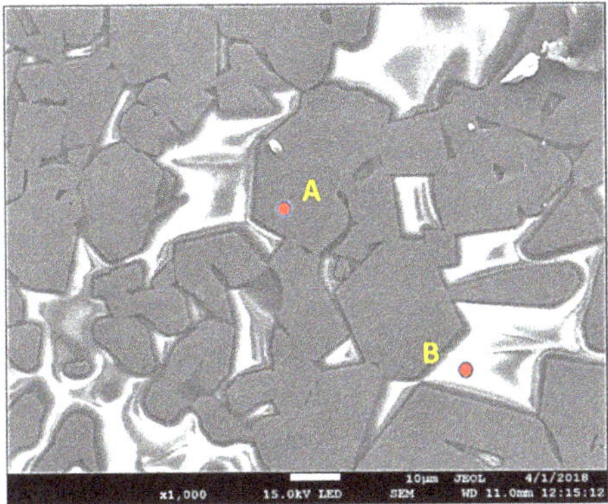

Figure 11. SEM-EDS of sinter sample adding 5% B_2O_3.

4. Conclusions

This work studied the effect of B_2O_3 on the properties of sinter produced in an 85 kg sinter test pot using 80% vanadium–titanium magnet concentrates (80 wt. %) and Australian hematite (20 wt. %). Analytical B_2O_3 was added to the sinter mix using mass ratios of 0.0–5.0% in the tests. The resulting sinter was analyzed to study the effect of boron addition. The following conclusions were drawn:

1. With B_2O_3 addition, the mechanical strength of the sinter first increased and then decreased, showing a maximum drum strength of 58% at 3.0% addition.
2. B_2O_3 addition significantly improved the low-temperature reduction pulverization of the sinter. However, when more than 1.0% was added, the improvement was limited. A low ratio of B_2O_3 was found to improve the reduction degree of the vanadium–titanium sinter, but large additions strongly reduced the reduction degree.
3. B_2O_3 addition yielded lower softening start and end temperatures of the sinter, and the softening interval grew significantly.
4. B_2O_3 addition changed the microstructure of the sinter from a large plate-like structure to a particle- and point-like structure. The total porosity was small, the diameter of the pores was large, and the number of cracks decreased.

In conclusion, the addition of boron oxide to the sinter mixture gives rise to both positive and negative effects for the resulting sinter consisting of vanadium–titanium-containing iron ores. To achieve the best performance of the sinter in the blast furnace, the addition of B_2O_3 should be limited to about 1%, so as to avoid the negative consequences of phases that melt at too low a temperature and within too wide a temperature interval.

Author Contributions: H.L., K.Z., X.X. contributed to perform the experiments, material characterization, data analysis and paper writing; H.S., W.L. and Y.Q. revised the paper and refined the language; and H.L. and Y.Q. contributed to the design of the experiment. All authors have read and agreed to the published version of the manuscript.

Funding: This work was supported by the Natural Science Foundation of Chongqing. (No. cstc2020jcyj-msxmX0583); The National Natural Science Foundation of China Grant No. 51974054); the Youth project of science and technology research program of Chongqing Education Commission of China. (No. KJQN201901); China Scholarship Council (201802075005).

Conflicts of Interest: The authors declare no conflict of interest.

References

1. Tang, W.; Yang, S.; Cheng, G.; Gao, Z.; Yang, H.; Xue, X. Effect of TiO_2 on the Sintering Behavior of Chromium-Bearing Vanadium–Titanium Magnetite. *Minerals* **2018**, *8*, 263. [CrossRef]

2. Qie, Y.N.; Lv, Q.; Zhang, X.S.; Liu, X.J.; Sun, Y.P. Effect of TiO_2 content on microstructure of sinter. *J. Iron Steel Res.* **2015**, *27*, 21–25.

3. Bristow, N.J.; Loo, C.E. Sintering Properties of Iron Ore Mixes Containing Titanium. *ISIJ Int.* **1992**, *32*, 819–828. [CrossRef]

4. Zhou, M.; Yang, S.T.; Jiang, T. Influence of basicity on high-chromium vanadium–titanium magnetite sinter properties productivity and mineralogy. *JOM* **2015**, *67*, 1203–1213. [CrossRef]

5. Cores, A.; Babich, A.; Muñiz, M.; Ferreira, S.; Mochon, J. The Influence of Different Iron Ores Mixtures Composition on the Quality of Sinter. *ISIJ Int.* **2010**, *50*, 1089–1098. [CrossRef]

6. Lv, X.W.; Bai, C.G.; Qiu, G.B. Moisture capacity: Definition, measurement, and application in determining the optimal water content in granulating. *ISIJ Int.* **2010**, *50*, 695–701. [CrossRef]

7. Kasai, E.; Sakano, Y.; Kawaguchi, T.; Nakamura, T. Influence of Properties of Fluxing Materials on the Flow of MeltFormed in the Sintering Process. *ISIJ Int.* **2000**, *40*, 857–862. [CrossRef]

8. Zhou, M.; Jiang, T.; Yang, S.T.; Xue, X. Vanadium–titanium magnetite ore blend optimization for sinter strength based on iron ore basic sintering characteristics. *Int. J. Min. Proces.* **2015**, *142*, 125–133. [CrossRef]

9. Shapovalov, A.N.; Ovchinnikova, E.V.; Gorbunov, V.B.; Dema, R.R.; Kalugina, O.B. The effect of the composition of magnesia flux on the sinter structure and properties. *IOP Conf. Ser. Mater. Sci. Eng.* **2019**, *625*, 012009. [CrossRef]

10. Umadevi, T.; Nelson, K.; Mahapatra, P.C.; Prabhu, M.; Ranjan, M. Influence of magnesia on iron ore sinter properties and productivity. *Ironmak. Steelmak.* **2009**, *36*, 515–520. [CrossRef]

11. Gao, Q.J.; Wei, G.; Shen, T.S.; Jiang, X.; Zheng, H.Y.; Shen, F.M.; Liu, C.S. Influence and mechanism of Indonesia vanadium titano-magnetite on metallurgical properties of iron ore sinter. *J. Cent. South Univ.* **2017**, *24*, 2805–2812. [CrossRef]

12. Yang, S.T.; Zhou, M.; Jiang, T. Effect of sintering essential characteristics on reduction degradation index for Cr-bearing vanadium and titanium magnetite. *J. Northeast. Univ. (Nat. Sci.)* **2015**, *36*, 498–501.

13. Zhou, M.; Yang, S.T.; Jiang, T. Effect of enhanced granulation on Cr-bearing vanadium and titanium magnetite sintering. *Iron Steel* **2015**, *50*, 39–43.

14. Higuchi, K.; Okazaki, J.; Nomura, S. Influence of Melting Characteristics of Iron Ores on Strength of Sintered Ores. *ISIJ Int.* **2020**, *60*, 674–681. [CrossRef]

15. Harvey, T.; Honeyands, T.; O'dea, D.; Evans, G. Study of Sinter Strength and Pore Structure Development usingAnalogue Tests. *ISIJ Int.* **2020**, *60*, 73–83. [CrossRef]

16. Chu, M.S.; Tang, Y. Improving metallurgical properties of high chromium vanadium titanium sinter with bornite. *J. Northeast. Univ. (Nat. Sci.)* **2015**, *12*, 22–25.

17. Hao, D.S. Study on the Experiment and Mechanism of Boracic Complex Additive to the Sinter. Ph.D. Thesis, Northeastern University, Shenyang, China, 2008.

18. Zhao, Q.J.; He, C.Q.; Gao, M.H. Reasonable utilization of boronic magnetite ore. *J.Anhui Univ. Technol. (Nat. Sci.)* **1997**, *3*, 262–266.

19. Xu, J.C.; Pang, Q.H.; Me, J.W.; Zhong, Q.; Yuan, P.; Tian, C. Influence of B on the microstructure of vanadium titanium iron ore sintering. *J. Univ. Sci. Technol. Liaoning* **2016**, *39*, 401–405.

20. Ren, S.; Zhang, J.L.; Xing, X.D.; Su, B.X. Effect of B_2O_3 on phase compositions of high Ti bearing titanomagnetite sinter. *Ironmak. Steelmak.* **2014**, *41*, 500–506. [CrossRef]

21. Ren, S.; Zhang, J.L.; Liu, Q.C.; Chen, M.; Ma, X.D.; Li, K.J.; Zhao, B.J. Effect of B_2O_3 on the reduction of FeO in Ti bearing blast furnace primary slag. *Ironmak. Steelmak.* **2015**, *42*, 498–503. [CrossRef]

22. Zhang, Y.Z. Behaviour and Action Law of Boron-Magnesia Powder Agglomeration. Ph.D. Thesis, Yanshan University, Qinhuangdao, China, 2002.

23. Liu, H.; Zhang, K.; Ling, Q.F.; Qin, Y.L. Effect of B_2O_3 Content on the Sintering Basic Characteristics of Mixed Ore Powder of Vanadium–Titanium Magnetite and Hematite. *J. Chem.* **2020**, *2020*, 6279176. [CrossRef]

24. GB/T 13241-2017. *Iron Ores-Determination of Reducibility*; General Administration of Quality Supervision, Inspection and Quarantine of the People's Republic of China: Beijng, China, 2017.

25. GB/T 13242-2017. *Iron Ores-Low-Temperature Disintegration Test-Method Using Cold Tumbling after Static Reduction*; General Administration of Quality Supervision, Inspection and Quarantine of the People's Republic of China: Beijng, China, 2017.

26. GB/T 34211-2017. *Method for Determination of Iron Reduction Softening Dripping Performance under Load*; General Administration of Quality Supervision, Inspection and Quarantine of the People's Republic of China: Beijng, China, 2017.

27. Zhang, K.; Yang, Z.X. Compressive strength and reducibility of B_2O_3-MgO-SiO_2-CaO-Fe_3O_4 Agglomerate. *J. Northeast. Univ. (Nat. Sci.)* **1997**, *6*, 589–592.

28. Shigaki, I.; Sawada, M.; Maekawa, M.; Narita, K. Fundamental study of size degradation mechanism of agglomerates during reduction. *Trans. ISIJ* **2006**, *22*, 838–847. [CrossRef]

Article

Distribution Behavior of Phosphorus in $2CaO \cdot SiO_2$-$3CaO \cdot P_2O_5$ Solid Solution Phase and Liquid Slag Phase

Bin Zhu [1,2], Mingmei Zhu [1,3,*], Jie Luo [1], Xiaofei Dou [1], Yu Wang [1,3], Haijun Jiang [2] and Bing Xie [1,3]

[1] College of Materials Science and Engineering, Chongqing University, Chongqing 400044, China; zhubincq@foxmail.com (B.Z.); 201909131200@cqu.edu.cn (J.L.); dxf09381@126.com (X.D.); wangyu@cqu.edu.cn (Y.W.); bingxie@cqu.edu.cn (B.X.)
[2] Chongqing Iron and Steel Research Institute Co. LTD, Chongqing 400084, China; 13627687756@139.com
[3] Chongqing Key Laboratory of Vanadium-Titanium Metallurgy and New Materials, Chongqing University, Chongqing 400044, China
* Correspondence: zhumingmei@cqu.edu.cn; Tel.: +86-23-6512-7306

Received: 28 July 2020; Accepted: 14 August 2020; Published: 17 August 2020

Abstract: In this paper, the CaO-SiO_2-Fe_tO-P_2O_5 dephosphorization slag system during the premier and middle stage of the converter process was studied, the effect of slag composition on the distribution ratio and activity coefficient of P in the $n \cdot 2CaO \cdot SiO_2$-$3CaO \cdot P_2O_5$ (recorded as nC_2S-C_3P) solid solution phase and liquid slag phase in the slag system was studied used by the high temperature experiment in laboratory, the theoretical calculation of thermodynamics, and the scanning electron microscope and the energy dispersive spectrometer (recorded as SEM/EDS). The research results show that when the FeO content in the liquid slag increases from 32.21% to 50.31%, the distribution ratio of phosphorus (recorded as L_P) in the liquid slag phase increases by 3.34 times. When the binary basicity in the liquid slag increases from 1.08 to 1.64, the L_P in the liquid slag phase decreases by 94.21%. In the initial slag, when the binary basicity increases from 2.0 to 3.5, the L_P decreases by 70.07%. When FeO content increases from 38.00% to 51.92%, the L_P increases by 6.15 times. When P_2O_5 content increases from 3.00% to 9.00%, the L_P increased by 10.67 times. When the FeO content in the liquid slag increases from 32.21% to 50.31%, the activity coefficient of P_2O_5 in the liquid slag phase (recorded as $\gamma_{P_2O_5(L)}$) increases by 54.33 times. When the binary basicity in the liquid slag increases from 1.08 to 1.64, $\gamma_{P_2O_5(L)}$ decreases by 99.38%. When the binary basicity increases from 2.0 to 3.5, the activity coefficient of P_2O_5 in the solid solution phase (recorded as $\gamma_{P_2O_5(SS)}$) in the solid solution phase decreases by 98.85%. When P_2O_5 content increases from 3.00% to 9.00%, $\gamma_{P_2O_5(SS)}$ increases by 1.14 times. When the binary basicity decreases from 3.5 to 2.0, n decreases from 0.438 to 0.404. When the FeO content increases from 38.00% to 51.92%, n decreases from 0.477 to 0.319. When the P_2O_5 content increases from 3.00% to 9.00%, n decreases from 0.432 to 0.164. The decrease of binary basicity and the increase of FeO and P_2O_5 content in the initial slag can reduce the value of n and enrich more phosphorus in the solid solution phase. The results can not only provide a theoretical basis for industrial production, but also lay a theoretical foundation for finding more effective dephosphorization methods.

Keywords: CaO-SiO_2-Fe_tO-P_2O_5 slag system; distribution ratio of phosphorus; dephosphorization; $n \cdot 2CaO \cdot SiO_2$-$3CaO \cdot P_2O_5$ solid solution

---

1. Introduction

Dephosphorization is one of the main tasks in the steelmaking process, and it is also a significant problem for integrated steelmakers. Phosphorus removal is required because phosphorus has a deleterious effect on the mechanical properties of steel. With the gradual depletion of high grade and low phosphorus iron ores resources in the world, the proportion of high phosphorus iron ore is increasing. Therefore, the requirements for the dephosphorization capacity of steelmaking slag system will be more stringent. Many studies have studied the distribution behavior of phosphorus between molten slag and molten steel [1–8], but the research on the distribution behavior of phosphorus between the solid solution phase and the liquid slag phase is not thorough enough.

The dephosphorization slag system of the converter steelmaking process mainly consists of the CaO-SiO_2-Fe_tO-P_2O_5 slag system, which usually exists in the state of a solid solution within the composition range of $2CaO \cdot SiO_2$ (recorded as C_2S) saturation in the slag system [9]. Fix et al. [10] plotted the pseudo binary phase diagram of the $2CaO \cdot SiO_2$-$3CaO \cdot P_2O_5$ (recorded as C_2S-C_3P) solid solution. The phase diagram shows that C_2S and $3CaO \cdot P_2O_5$ (recorded as C_3P) can form a stable solid solution in any composition range during the temperature range of the steelmaking process. Some studies [11–15] showed that phosphorus in the CaO-SiO_2-Fe_tO-P_2O_5 slag system mainly exists in the form of the nC_2S-C_3P solid solution, which is irregular granular or a long strip in slag, while almost no phosphorus or a small amount of phosphorus is in other phases. When the composition and temperature of the slag are in a certain range, phosphorus will exist in the independent form of C_3P [14]. The pseudo binary phase diagram of C_2S-C_3P plotted by Fix et al. [10] and the research results of Kitamura et al. [16] showed that phosphorus is very stable in the C_2S-C_3P solid solution in the steelmaking temperature range. All of the above results indicate that during the steelmaking process of the converter, phosphorus can be migrated as much as possible by changing the composition of slag and be stably present in the solid solution, which is beneficial to reduce the concentration of phosphorus in the liquid slag phase, improve the driving force of mass transfer of phosphorus from molten iron to the liquid slag phase, and improve the dephosphorization efficiency of converter steelmaking. Therefore, the C_2S-C_3P solid solution plays a key role in the dephosphorization process of converter steelmaking. It is necessary to study the influence of slag composition on the distribution ratio of phosphorus between the nC_2S-C_3P solid solution and liquid slag.

Some researchers have studied the effect of C_2S in the dephosphorization slag system on dephosphorization efficiency. Suito et al. [17] found that according to the experimental study, the dephosphorization rate would be greatly improved when the slag system contains more solid lime particles and dicalcium silicate particles. Shimauchi et al. [14] studied the effect of slag composition on the distribution ratio of phosphorus between the solid solution phase and the liquid slag. The results showed that the higher the content of T. Fe in slag, the higher the content of phosphorus in the solid solution. Son et al. [18] analyzed the effect of binary basicity and FeO content of slag on the precipitation of calcium silicate in situ by hot thermocouple technology (HTT). The results showed that the distribution ratio of phosphorus increased significantly when there was solid solution precipitation. In the current research work, there is little research on the distribution behavior of phosphorus in the solid and liquid phases of nC_2S-C_3P in the dephosphorization slag system of the converter. Therefore, it is necessary to systematically study the influence of slag composition on the distribution behavior of phosphorus in the solid and liquid phases.

In this paper, the dephosphorization slag system of CaO-SiO_2-Fe_tO-P_2O_5 during the converter steelmaking process is taken as the research object. By means of a high temperature experiment and SEM/EDS detection, the distribution ratio of the slag composition to phosphorus in the nC_2S-C_3P solid solution phase and liquid slag phase is systematically studied. The influence of slag composition on the activity coefficient of phosphorus in the solid solution phase and liquid slag phase is studied by means of thermodynamics theory analysis. The research results can not only provide a theoretical basis for the establishment of the slagging system in the converter steelmaking process, but also lay a theoretical foundation for finding more effective dephosphorization methods.

2. Experiment

The dephosphorization slag system in the premier and middle stages of converter steelmaking process has the characteristics of high basicity and strong oxidation. At this time, the temperature in the converter is relatively low, and the dephosphorization capacity of the slag system is also high under the condition that the slag has good fluidity. According to the characteristics of the slag system studied, the slag was prepared in the laboratory according to the composition shown in Table 1. The binary basicity of the slag system studied is in the range of 2.0–3.5, the content of FeO is 38.00–51.98%, and the content of P_2O_5 is in the range of 3.00–9.00%. According to the position in the liquidus diagram of the CaO-FeO-SiO$_2$ ternary system, the slag compositions are all in the C$_2$S primary crystal zone to ensure that the calcium silicate precipitated from the slag during cooling process are all C$_2$S, and the compositions are basically between 1400 and 1500 °C liquidus, as shown in Figure 1. High temperature silicon molybdenum furnace was used in the experiment, and the temperature control precision was ±2 °C. In the experiment, the slag samples were placed in the MgO crucible, and the MgO crucible was covered with graphite crucible. During the experiment, argon gas was introduced as protective gas. The temperature control curve during the experiment is shown in Figure 2. When the temperature rises to 1500 °C, the slag is kept for 1–2 h to make the slag melt evenly; in order to precipitate more phosphorus rich phases, it is reduced to 1400 °C at a slow rate of 1.67 K/min and kept for 1 h; when the precipitates and liquid phase are uniform, take out the crucible immediately and quickly cool it with liquid nitrogen. The slag samples were taken out, ground, and polished with sandpaper after drying. Then, the samples were detected and analyzed by SEM/EDS (Carl Zeiss AG, Oberkochen, Germany) after spraying gold. When analyzing the content of each composition in each phase by EDS (Carl Zeiss AG, Oberkochen, Germany), the method of taking the average value for many times is adopted to reduce the error caused by EDS detection as far as possible. The experiment did not achieve the equilibrium state, and the experimental data obtained based on the high temperature experiment in this article are all the experimental results of the non-equilibrium state.

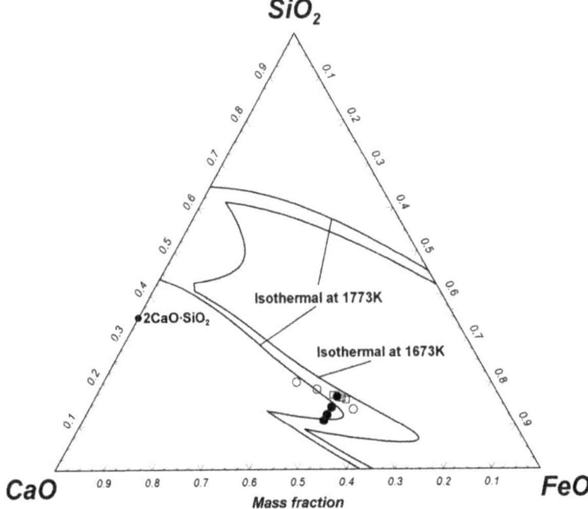

Figure 1. Position of the slag composition in the liquidus diagram of the CaO-FeO-SiO$_2$ ternary system at 1400 °C and 1500 °C.

Table 1. Composition of the slag samples/wt %.

No.	(%CaO)/(%SiO$_2$)	CaO	SiO$_2$	FeO	P$_2$O$_5$
1	2.0	31.67	15.83	47.50	5.00
2	2.5	33.93	13.57	47.50	5.00
3	3.0	35.63	11.88	47.50	5.00
4	3.5	36.94	10.56	47.50	5.00
5	2.0	38.00	19.00	38.00	5.00
6	2.0	34.83	17.42	42.75	5.00
7	2.0	28.68	14.34	51.98	5.00
8	2.0	33.00	16.50	47.50	3.00
9	2.0	30.33	15.17	47.50	7.00
10	2.0	29.00	14.50	47.50	9.00

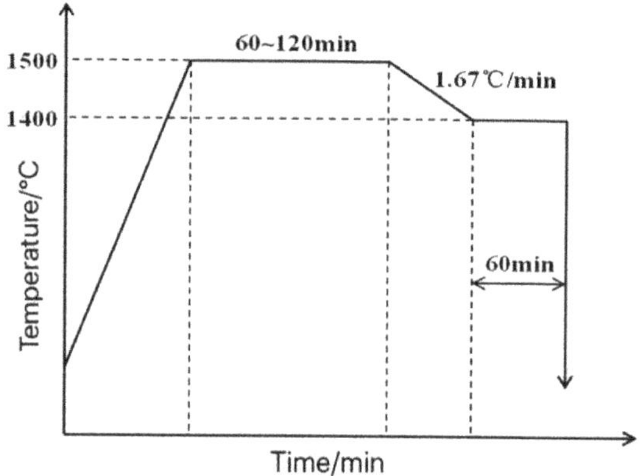

Figure 2. Temperature control curve.

3. Results

3.1. Definition of the Phosphorus Distribution Ratio and Phosphorus Activity Coefficient

The distribution ratio of phosphorus between the solid solution phase and the liquid slag phase, L_P, can characterize the dephosphorization effect of the slag systems. The larger the L_P, the more phosphorus in the solid solution phase; conversely, the more phosphorus in the liquid slag phase. Since the phosphorus in the liquid slag phase may return to the molten steel in the middle and final stage of converter process, the less phosphorus in the liquid slag phase, that is to say, the greater L_P, the better dephosphorization effect of slag systems.

When the L_P of the slag sample is analyzed by SEM, L_P is defined as shown in Equation (1).

$$L_P = A_{SS}(\%P_2O_5)_{SS}/A_L(\%P_2O_5)_L \tag{1}$$

Among them, A represents the area fraction of solid solution phase or liquid slag phase, %, and the area fraction A is calculated by Image Pro Plus software; the subscripts SS and L represent the solid solution phase and liquid slag phase of nC$_2$S-C$_3$P respectively, the same below; (%P$_2$O$_5$) represents the mass percentage content of P$_2$O$_5$.

When L_P is defined by the thermodynamic analysis, based on the regular solution model [18], the activity of P_2O_5 in the CaO-SiO_2-Fe_tO-P_2O_5 slag system was calculated, as shown in Equations (2) and (3).

$$RT \ln a_{P_2O_5(l)} = RT \ln a_{P_2O_5(RS)} + 52720 - 230.706T = 2RT \ln \gamma_{PO_{2.5}(RS)}$$
$$+RT \ln x_{P_2O_5(RS)} + 52720 - 230.706T \, (J) \tag{2}$$

$$RT \ln \gamma_{PO_{2.5}(RS)} = -251040x_{CaO}^2 + 83680x_{SiO_2}^2 - 37660x_{MgO}^2 - 31380x_{FeO}^2 -$$
$$33470x_{CaO}x_{SiO_2} - 188280x_{CaO}x_{MgO} - 243930 - 251040x_{CaO}x_{FeO} + \tag{3}$$
$$112960x_{SiO_2}x_{MgO} + 94140x_{SiO_2}x_{FeO} - 102510x_{MgO}x_{FeO} \, (J)$$

Among them, the subscript RS is the regular solution; γ is the activity coefficient; a is the activity; x is the molar percentage content, %; and T is the Kelvin temperature, K.

In the CaO-SiO_2-Fe_tO-P_2O_5 slag system, when the liquid slag phase and nC_2S-C_3P solid solution phase are in equilibrium, the activity of P_2O_5 in these two phases should be equal, that is, as shown in Equation (4).

$$a_{P_2O_5(SS)} = a_{P_2O_5(L)} \tag{4}$$

Therefore, the phosphorus distribution ratio can be written as Equation (5).

$$L_P = \frac{(\%P_2O_5)_{SS}}{(\%P_2O_5)_L} = k\frac{a_{P_2O_5(SS)} \times \gamma_{P_2O_5(L)}}{a_{P_2O_5(L)} \times \gamma_{P_2O_5(SS)}} = k\frac{\gamma_{P_2O_5(L)}}{\gamma_{P_2O_5(SS)}} \tag{5}$$

According to this, the activity coefficient of P_2O_5 in the nC_2S-C_3P solid solution phase can be obtained, as shown in Equation (6).

$$\gamma_{P_2O_5(SS)} = k\frac{\gamma_{P_2O_5(L)}}{L_P} \tag{6}$$

Among them, k is the conversion coefficient between the mole fraction and the mass fraction of P_2O_5 in the solid solution phase and the liquid slag phase, as shown in Equation (7).

$$k = \frac{(\%P_2O_5)_{SS} \times (x_{P_2O_5})_L}{(\%P_2O_5)_L \times (x_{P_2O_5})_{SS}} \tag{7}$$

3.2. Test and Calculation Results

Taking samples No. 8 and No. 10 as examples, the SEM photographs of slag samples are shown in Figure 3. Taking samples No. 8 and No. 10 as examples, the SEM photographs of slag samples are shown in Figure 3. The irregular black particles or long strips are the nC_2S-C_3P solid solution phase, and the gray or light gray continuous distribution areas are the liquid slag phase, and the white area is the RO phase. RO phase is the general term of the solid solutions formed by the divalent metal oxides such as FeO, MgO and MnO in the slags. No MgO was added to the experimental slag. The MgO detected by EDS originated from the MgO crucible used in the experiment. The MgO crucible was eroded by the slag at high temperature and entered the slag. Therefore, there is little RO phase in the slag, which is different from the slag sample in the practical production. The EDS detection results of each phase in the 10 schemes were converted into oxides of each element, as shown in Table 2. It can be seen from the table that most of the phosphorus exists in the solid solution phase.

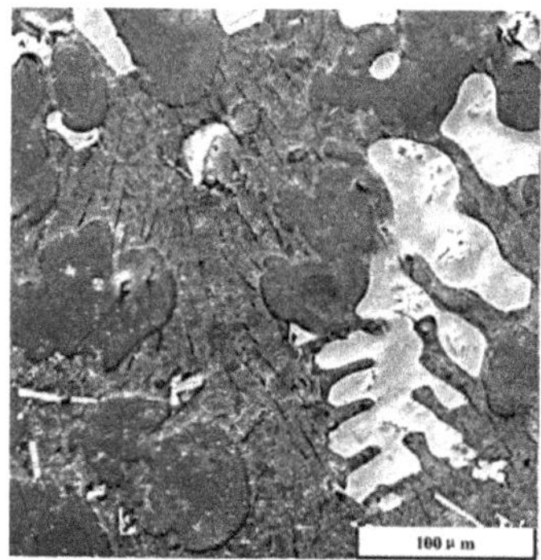 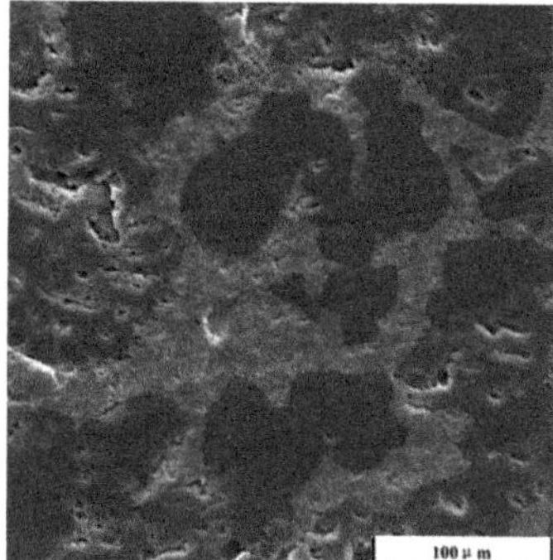

(**a**) Sample No. 8. (**b**) Sample No. 10.

Figure 3. Photographs of samples.

Table 2. Composition of the solid solution phase and liquid slag phase/wt %.

No.	Phase	CaO	SiO$_2$	FeO	MgO	P$_2$O$_5$	A/%	L_P
1	Solid solution phase	58.12	23.75	5.08	0.56	12.50	46.97	9.22
	Liquid slag phase	28.64	20.5	47.2	2.46	1.20	53.03	
2	Solid solution phase	58.88	25.57	1.82	1.00	12.73	49.98	5.63
	Liquid slag phase	31.21	24.36	40.93	1.24	2.26	50.02	
3	Solid solution phase	61.06	23.04	3.27	1.08	11.55	56.39	3.44
	Liquid slag phase	34.08	24.51	35.16	1.91	4.34	43.61	
4	Solid solution phase	64	21.11	4.22	0.85	9.82	58.55	2.76
	Liquid slag phase	36.42	24.98	32.21	1.36	5.03	41.45	
5	Solid solution phase	62.87	26.39	2.39	1.06	7.29	-	1.50
	Liquid slag phase	36.24	22.15	34.23	2.53	4.85	-	
6	Solid solution phase	59.70	25.21	4.27	1.13	9.69	42.78	2.93
	Liquid slag phase	31.21	21.12	42.23	2.97	2.47	57.22	
7	Solid solution phase	57.42	17.95	2.57	1.23	20.83	42.00	9.92
	Liquid slag phase	25.77	19.35	50.31	3.05	1.52	58.00	
8	Solid solution phase	59.68	24.96	4.22	0.60	10.54	33.80	2.43
	Liquid slag phase	29.23	19.22	45.9	2.71	2.94	49.83	
	RO phase	3.01	-	78.19	18.8	-	16.37	
9	Solid solution phase	54.36	15.48	5.06	1.07	24.03	58.18	15.69
	Liquid slag phase	25.86	21.34	49.49	1.18	2.13	41.82	
10	Solid solution phase	51.41	9.69	4.79	0.79	33.32	68.84	25.92
	Liquid slag phase	23.9	22.05	50.27	1.24	2.84	31.16	

3.3. Composition Distribution Characteristics of the Solid Solution Phase and Liquid Slag Phase

According to the composition of CaO, SiO$_2$, and P$_2$O$_5$ in the solid solution phase and the liquid slag phase obtained in Table 2, the composition of the solid solution phase and the liquid slag phase of the ten slag samples were characterized in the phase diagram of the CaO-SiO$_2$-P$_2$O$_5$ ternary system. The positions of the solid solution phase and the liquid slag phase in the slag samples in the phase diagram are shown in Figures 4 and 5, respectively. It can be seen that the solid solution phase compositions were basically on the connection line of C$_2$S and C$_3$P. It indicates that the solid solution was composed of C$_2$S and C$_3$P to form a pseudo binary relationship, and the solid solution was formed by combining C$_2$S and C$_3$P in different proportions. The composition of the liquid slag phase was basically near the liquidus at 1400 °C, which indicates that the slag sample could basically represent the state near 1400 °C.

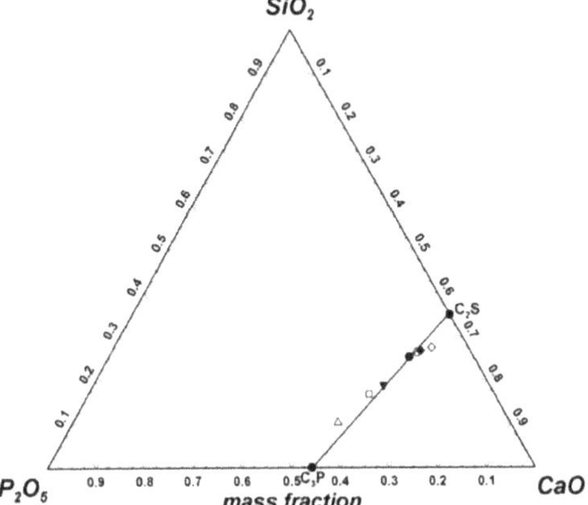

Figure 4. Position of solid solution phase components in different slag samples in the phase diagram of the CaO-SiO$_2$-P$_2$O$_5$ ternary system.

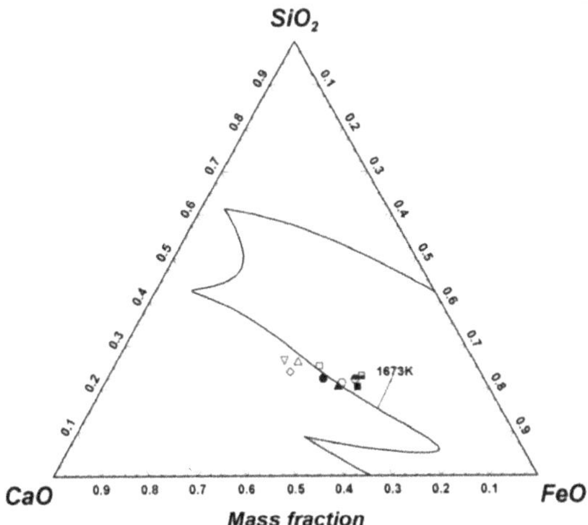

Figure 5. Position of liquid slag phase components in different slag samples in the phase diagram of the CaO-SiO$_2$-FeO ternary system.

4. Discussions

4.1. Effect of Slag Compositions on L_P

Figure 6a,b shows the effects of FeO content and binary basicity in the liquid slag phase on L_P, respectively. When the FeO content in the liquid slag increased from 32.21% to 50.31%, the L_P increased from 2.76 to 9.22, and the L_P in the liquid slag phase increased by 3.34 times. When the binary basicity in the liquid slag increased from 1.08 to 1.64, the L_P decreased from 25.92 to 1.50, and the L_P in the liquid slag phase decreased by 94.21%. The effect of FeO and binary basicity in liquid slag on L_P was just the opposite. This is because the increase of FeO content in liquid slag can reduce the viscosity of the molten slag, and it is beneficial to help the phosphorus in liquid slag diffuse into the solid solution phase, thus increasing the distribution ratio of phosphorus in the solid solution phase and liquid slag phase. The increase of binary basicity in liquid slag will increase the viscosity of molten slag, which ultimately leads to the decrease in the distribution ratio of phosphorus between the solid solution phase and the liquid slag phase. The influence trend of FeO content in the liquid slag phase on L_P was consistent with the research results of reference [9].

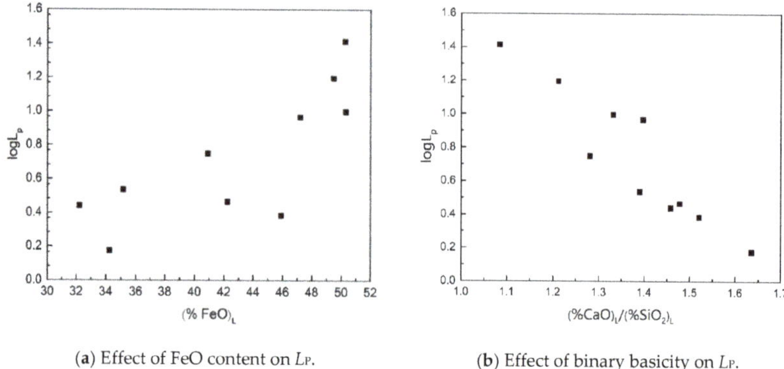

(a) Effect of FeO content on L_P. (b) Effect of binary basicity on L_P.

Figure 6. Effect of FeO content and binary basicity in the liquid slag phase on L_P.

Figure 7a–c shows the effect of the compositions of the experimental primary slag on the L_P. Essentially, the effect of the compositions of the experimental primary slag on L_P is only apparent, but the essence is the influence of the compositions of the liquid slag and solid solution phase caused by the compositions of the initial slag on the L_P. It can be seen from Figure 7 that within the range of the initial slag composition, when the binary basicity increased from 2.0 to 3.5, the L_P decreased from 9.22 to 2.76, and the L_P decreased by 70.07%. When FeO content increased from 38.00% to 51.92%, the L_P increased from 1.50 to 9.22, and the L_P increased by 6.15 times. When P_2O_5 content increased from 3.00% to 9.00%, the L_P increased from 2.43 to 25.92, and the L_P increased by 10.67 times. This is because when the binary basicity of the experimental initial slag is greater than 2.0, with the increase of basicity, the CaO content in the liquid slag phase will increase, which will lead to the decrease of L_P. The increase of FeO content in the experimental initial slag will cause the increase of FeO content in the liquid slag phase, which will cause the increase of L_P. The increase of P_2O_5 content increased the concentration difference of phosphorus diffusion from the liquid slag phase to the solid solution phase, and increased the driving force of phosphorus migration from the liquid slag phase to the solid solution phase, thus promoting the increase of L_P.

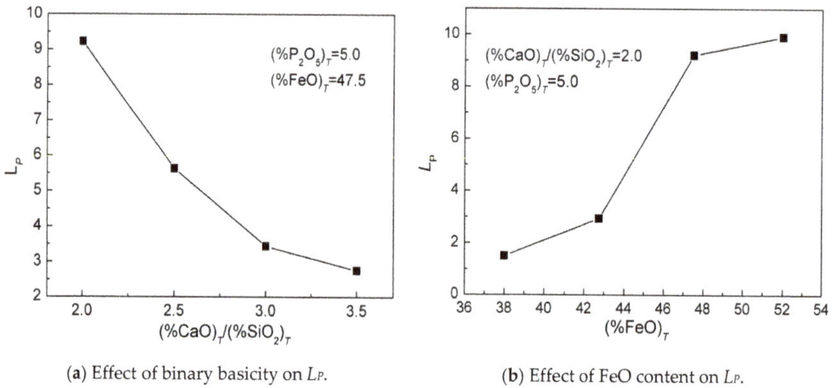

(a) Effect of binary basicity on L_P. (b) Effect of FeO content on L_P.

Figure 7. *Cont.*

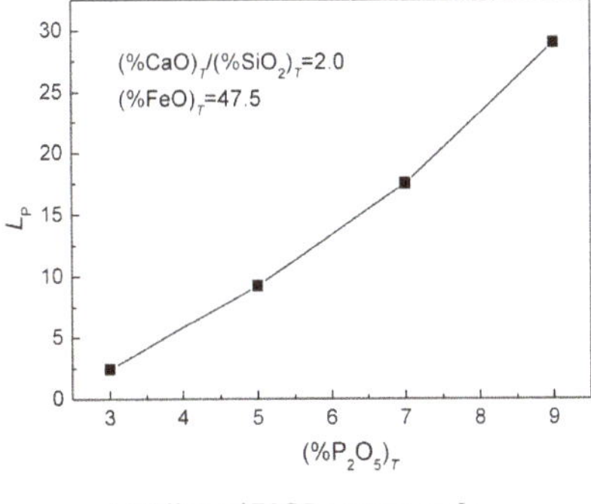

(**c**) Effect of P2O5 content on L_P.

Figure 7. Effect of the components of the initial slag on L_P. Note: The subscript t represents the experimental initial slag, which is the same below.

The slag compositions of the 10 groups of schemes and their corresponding L_P values were plotted, and Figure 8 was obtained. Figure 8 shows the slag composition and its corresponding L_P values in the CaO-(SiO$_2$+ P$_2$O$_5$)-FeO pseudo ternary system phase diagram. It can be seen from the figure that L_P increased in the direction toward FeO and P$_2$O$_5$ content, which indicates that the increase of FeO and P$_2$O$_5$ content in the slag was beneficial to the diffusion of phosphorus from the liquid slag phase to the solid solution phase. L_P decreased in the direction toward CaO content, which indicates that when the binary basicity of the initial slag was greater than 2.0, the increase of CaO content in the slag was not beneficial to the diffusion of phosphorus from the liquid slag phase to the solid solution phase.

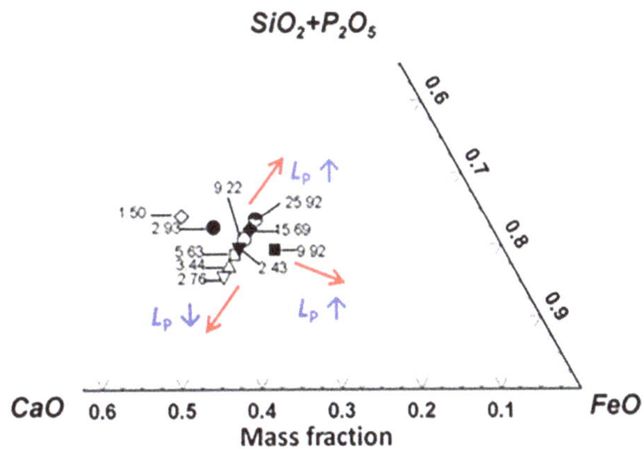

Figure 8. Slag composition and its corresponding L_P value in the CaO-(SiO$_2$ + P$_2$O$_5$)-FeO pseudo ternary composition diagram.

4.2. Effect of Slag Composition on the Activity Coefficient of Phosphorus

Figure 9 shows the effect of FeO content and binary basicity in the liquid slag phase on the activity coefficient of P$_2$O$_5$ calculated by the regular solution model. It can be seen from the figure that within the range of the initial slag composition, when the FeO content in the liquid slag increased from 32.21% to 50.31%, $\gamma_{P_2O_5(L)}$ increased from 2.54×10^{-15} to 1.38×10^{-13} and increased by 54.33 times. When the binary basicity in the liquid slag increased from 1.08 to 1.64, $\gamma_{P_2O_5(L)}$ decreased from 2.65×10^{-13} to 1.64×10^{-15} and decreased by 99.38%. The results show that the increase of FeO content could reduce the stability of phosphorus in the liquid slag phase, which is helpful for phosphorus diffusion from the liquid slag phase to the solid solution phase. While the increase of binary basicity could increase the stability of phosphorus in the liquid slag phase, which is not beneficial for the diffusion of phosphorus into the solid solution phase. According to the calculation results, $\gamma_{P_2O_5(L)}$ was mainly affected by the content of FeO and binary basicity in the liquid slag, and other components had little influence on it. Comparing Figure 9 with Figure 6, it can be seen that the effect of the liquid slag composition on the phosphorus activity coefficient calculated by regular solution model was consistent with the experimental results.

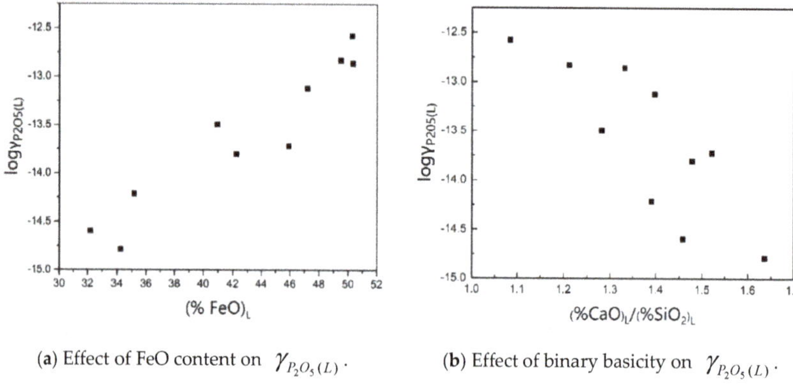

(**a**) Effect of FeO content on $\gamma_{P_2O_5(L)}$.   (**b**) Effect of binary basicity on $\gamma_{P_2O_5(L)}$.

Figure 9. Effect of FeO content and binary basicity in the liquid slag phase on $\gamma_{P_2O_5(L)}$.

Figures 10 and 11 show the effects of the binary basicity of the experimental initial slag and P_2O_5 content in the solid solution phase on the activity coefficient of P_2O_5 in the solid solution phase. It can be seen from the figure, when the binary basicity increased from 2.0 to 3.5, $\gamma_{P_2O_5(SS)}$ in the solid solution phase decreased from 8.34×10^{-14} to 9.56×10^{-16} and decreased by 98.85%. When P_2O_5 content increased from 3.00% to 9.00%, $\gamma_{P_2O_5(SS)}$ increased from 8.14×10^{-15} to 9.28×10^{-15} and increased by 1.14 times. It shows that the increase of the basicity in the initial slag was beneficial to improve the stability of the phosphorus in the solid solution phase. This is because under the experimental conditions, the increase of the basicity in the initial slag was helpful to the formation of the solid solution phase and the fixation of phosphorus. The increase of the P_2O_5 content in the solid solution phase will reduce its stability. This is because as the amount of phosphorus fixed in the solid solution phase increases, the phosphorus fixation capacity will decrease accordingly.

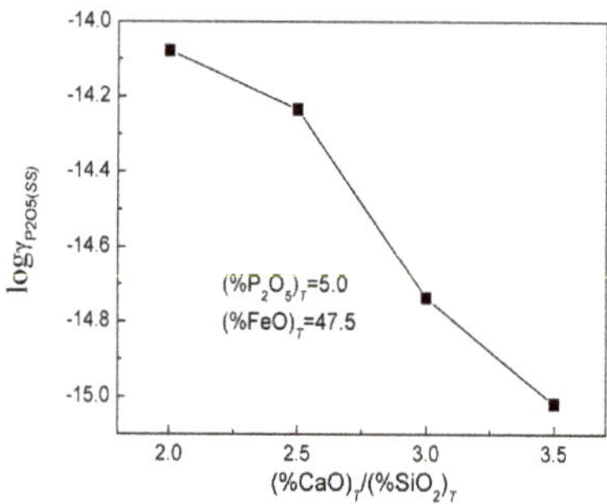

Figure 10. Effect of binary basicity in the initial slag on $\gamma_{P_2O_5(SS)}$.

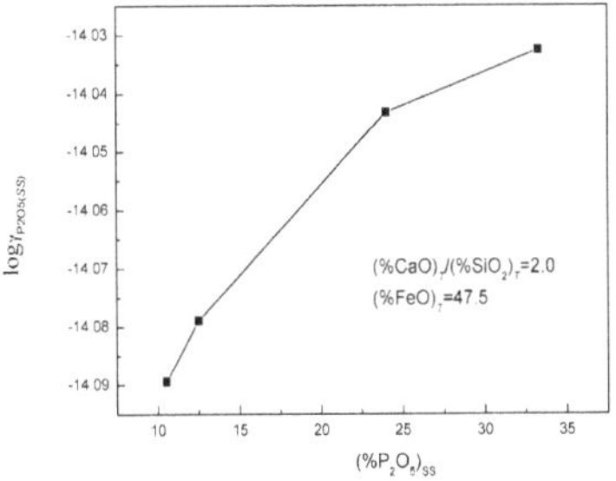

Figure 11. Effect of P_2O_5 content in the solid solution phase on $\gamma_{P_2O_5(SS)}$.

4.3. *n in the nC_2S-C_3P Solid Solution*

The ratio of C_2S and C_3P in the nC_2S-C_3P solid solution phase will change with different slag compositions, that is to say, the value of n will change. If the value of n in the nC_2S-C_3P solid solution molecular formula is smaller, the proportion of C_3P in the solid solution phase is larger. Therefore, the value of n can characterize the amount of phosphorus enriched in the solid solution phase. Figure 12 shows the influence of composition changes on the value of n in the solid solution phase within the range of the initial slag composition. It can be seen from the figure that within the range of the initial slag composition in the experiment, when the binary basicity decreased from 3.5 to 2.0, n decreased from 0.438 to 0.404 and decreased by 7.76%. When the FeO content increased from 38.00% to 51.92%, n decreased from 0.477 to 0.319 and decreased by 33.12%. When the P_2O_5 content increased from 3.00% to 9.00%, n decreased from 0.432 to 0.164 and decreased by 62.04%. The decrease of basicity and the increase of FeO and P_2O_5 content in the initial slag can reduce the value of n, that is, more phosphorus is enriched in the solid solution phase.

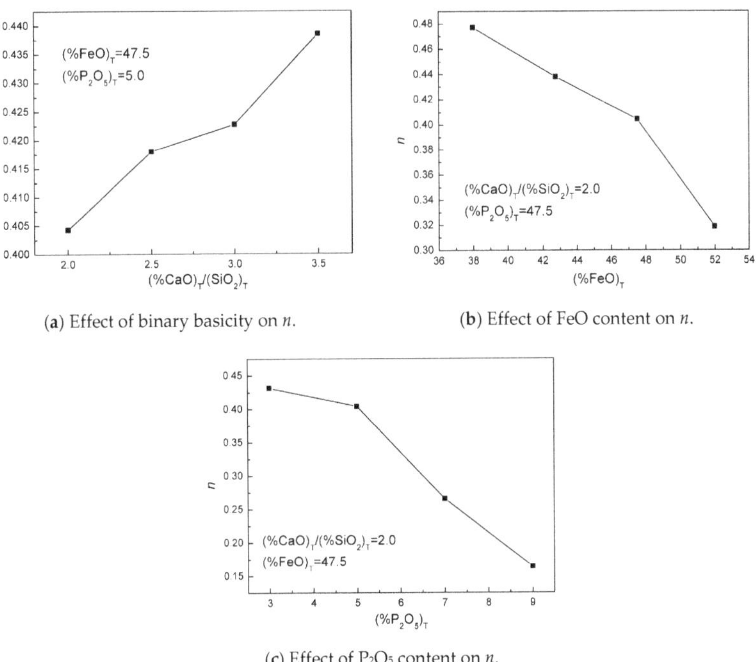

(**a**) Effect of binary basicity on n.

(**b**) Effect of FeO content on n.

(**c**) Effect of P_2O_5 content on n.

Figure 12. Effect of the slag compositions on n.

5. Conclusions

In this paper, the $CaO\text{-}SiO_2\text{-}Fe_tO\text{-}P_2O_5$ slag system was taken as the research object, and the combination method of high temperature experiments and thermodynamic calculations was used. The compositions of the $nC_2S\text{-}C_3P$ solid solution phase and the liquid slag phase in the $CaO\text{-}SiO_2\text{-}Fe_tO\text{-}P_2O_5$ slag system prepared in the laboratory were analyzed by means of SEM/EDS detection methods. The influence of the slag composition on the distribution ratio and activity coefficient of phosphorus between the $nC_2S\text{-}C_3P$ solid solution phase and the liquid slag phase was studied. The main conclusions are as follows:

(1) When the FeO content in the liquid slag increased from 32.21% to 50.31%, the L_P in the liquid slag phase increased by 3.34 times. When the binary basicity in the liquid slag increased from 1.08 to 1.64, the L_P in the liquid slag phase decreased by 94.21%.

(2) When the binary basicity increased from 2.0 to 3.5, the L_P decreased by 70.07%. When FeO content increased from 38.00% to 51.92%, the L_P increased by 6.15 times. When P_2O_5 content increased from 3.00% to 9.00%, the L_P increased by 10.67 times.

(3) When the FeO content in the liquid slag increased from 32.21% to 50.31%, $\gamma_{P_2O_5(L)}$ increased by 54.33 times. When the binary basicity in the liquid slag increased from 1.08 to 1.64, $\gamma_{P_2O_5(L)}$ decreased by 99.38%.

(4) When the binary basicity increased from 2.0 to 3.5, $\gamma_{P_2O_5(SS)}$ in the solid solution phase decreased by 98.85%. When P_2O_5 content increased from 3.00% to 9.00%, $\gamma_{P_2O_5(SS)}$ increased by 1.14 times.

(5) When the binary basicity decreased from 3.5 to 2.0, n decreased from 0.438 to 0.404. When the FeO content increased from 38.00% to 51.92%, n decreased from 0.477 to 0.319. When the P_2O_5 content increased from 3.00% to 9.00%, n decreased from 0.432 to 0.164. The decrease of basicity and the increase of FeO and P_2O_5 content in the initial slag can enrich more phosphorus in the solid solution phase.

Author Contributions: Methodology, Y.W.; formal analysis, J.L. and X.D.; investigation, X.D.; data curation, H.J. and X.D.; writing—original draft preparation, M.Z.; writing—review and editing, B.Z.; supervision, B.X. All authors have read and agreed to the published version of the manuscript.

Funding: This research received no external funding.

References

1. Assis, N.A.; Tayeb, M.A.; Sridhar, S.; Fruehan, R.J. Phosphorus equilibrium between liquid iron and $CaO\text{-}SiO_2\text{-}MgO\text{-}Al_2O_3\text{-}FeO\text{-}P_2O_5$ Slags: EAF slags, the effect of alumina and new correlation. *J. Met.* **2019**, *9*, 116.

2. Du, C.M.; Lv, N.N.; Su, C.; Liu, W.M.; Yang, J.X.; Wang, H.C. Distribution of P2O5 between solid solution and liquid phase in dephosphorization slag of CaO–SiO2–FeO–P2O5–Na2O system. *J. Iron Steel Res. Int.* **2019**, *26*, 1162–1170. [CrossRef]

3. Li, B.; Li, L.; Guo, H.; Guo, J.; Duan, S.; Sun, W. A phosphorus distribution prediction model for CaO-SiO2-MgO-FeO-Fe2O3-Al2O3-P2O5 slags based on the IMCT. *Ironmak. Steelmak.* **2019**, 1–10. [CrossRef]

4. Lin, L.; Bao, Y.P.; Gu, C.; Wu, W.; Zeng, J.Q. Distribution of P2O5 between P-rich phase and matrix phase in P-bearing steelmaking slag. *J. High Temp. Mater. Process.* **2018**, *37*, 655–663. [CrossRef]

5. Drain, P.B.; Monaghan, B.J.; Longbottom, R.J.; Chapman, M.W.; Zhang, G.; Chew, S.J. Phosphorus partition and phosphate capacity of basic oxygen steelmaking slags. *ISIJ Int.* **2018**, *58*, 1965–1971. [CrossRef]

6. Drain, P.B.; Monaghan, B.J.; Zhang, G.; Longbottom, R.J.; Chapman, M.W.; Chew, S.J. A review of phosphorus partition relations for use in basic oxygen steelmaking. *Ironmak. Steelmak.* **2017**, *44*, 721–731. [CrossRef]

7. Han, X.; Li, J.; Zhou, C.G.; Shi, C.B.; Zheng, D.L.; Zhang, Z.M. Study on mineralogical structure of dephosphorisation slag at the first deslagging in BOF steelmaking process. *Ironmak. Steelmak.* **2017**, *44*, 262–268. [CrossRef]

8. Assis, A.N.; Tayeb, M.A.; Sridhar, S.; Fruehan, R.J. Phosphorus equilibrium between liquid iron and CaO-SiO2-MgO-Al2O3-FeO-P2O5 slag part 1: Literature review, methodology, and BOF slags. *Met. Mater. Trans. B* **2015**, *46*, 2255–2263. [CrossRef]

9. Ito, K.; Yanagisawa, M.; Sano, N. Phosphorus distribution between solid $2CaO \cdot SiO_2$ and molten $CaO-SiO_2-FeO-Fe_2O_3$ Slags. *J. Tetsu Hagané* **1982**, *68*, 150–152.

10. Fix, W.; Heymann, H.; Heinke, R. Subsolidus relations in the system $2CaO \cdot SiO_2$-$3CaO \cdot P_2O_5$. *J. Am. Ceram. Soc.* **1969**, *52*, 342–346. [CrossRef]

11. Inoue, R.; Suito, H. Phosphorous partition between $2CaO \cdot SiO_2$ particles and $CaO-SiO_2-Fe_tO$ slags. *J. ISIJ Int.* **2006**, *46*, 174–179. [CrossRef]

12. Inoue, R.; Suito, H. Mechanism of dephosphorization with $CaO-SiO_2-Fe_tO$ slags containing mesoscopic scale $2CaO \cdot SiO_2$ particles. *J. ISIJ Int.* **2006**, *46*, 188–194. [CrossRef]

13. Fukagai, S.; Hamano, T.; Tsukihashi, F. Formation reaction of phosphate compound on multiphase flux at 1573K. *J. ISIJ Int.* **2007**, *47*, 187–189. [CrossRef]

14. Shimauchi, K.; Kitamura, S.; Shibata, H. Distribution of P_2O_5 between solid dicalcium silicate and liquid phases in $CaO-SiO_2-Fe_2O_3$ system. *J. ISIJ Int.* **2009**, *49*, 505–511. [CrossRef]

15. Pahlevani, F.; Kitamura, S.Y.; Shibata, H.; Maruoka, N. Distribution of P_2O_5 between solid solution of $2CaO \cdot SiO_2$ and $3CaO \cdot P_2O_5$ and liquid phase. *J. ISIJ Int.* **2010**, *50*, 822–829. [CrossRef]

16. Kitamura, S.Y.; Saito, S.; Utagawa, K.; Shibata, H.; Robertson, D.G. Mass transfer of P_2O_5 between liquid slag and solid solution of $2CaO \cdot SiO_2$ and $3CaO \cdot P_2O_5$. *J. ISIJ Int.* **2009**, *49*, 1838–1844. [CrossRef]

17. Inoue, R.; Suito, H. Behavior of phosphorus transfer from $CaO-Fe_tO-P_2O_5$ (-SiO_2) slag to CaO particles. *J. ISIJ Int.* **2006**, *46*, 180–187.

18. Son, P.K.; Kashiwaya, Y. Phosphorus partition in dephosphorization slag occurring with crystallization at intial stage of solidification. *J. ISIJ Int.* **2008**, *48*, 1165–1174. [CrossRef]

Article

Self-Reduction Behavior of Bio-Coal Containing Iron Ore Composites

Asmaa A. El-Tawil [1,*], Hesham M. Ahmed [1,2], Lena Sundqvist Ökvist [1,3] and Bo Björkman [1]

[1] Department of Civil, Environmental and Natural Resources Engineering, Minerals and Metallurgical Engineering, MiMeR, Luleå University of Technology, 97187 Luleå, Sweden; Hesham.ahmed@ltu.se (H.M.A.); lena.sundqvist-oqvist@ltu.se (L.S.Ö.); Bo.Bjorkman@ltu.se (B.B.)
[2] Central Metallurgical Research and Development Institute, P.O Box 87, Helwan, 11421 Cairo, Egypt
[3] Swerim AB, 971 25 Luleå, Sweden
* Correspondence: asmaa.el-tawil@ltu.se; Tel.: +46-92-049-3131

Received: 19 December 2019; Accepted: 14 January 2020; Published: 16 January 2020

Abstract: The utilization of CO_2 neutral carbon instead of fossil carbon is one way to mitigate CO_2 emissions in the steel industry. Using reactive reducing agent, e.g., bio-coal (pre-treated biomass) in iron ore composites for the blast furnace can also enhance the self-reduction. The current study aims at investigating the self-reduction behavior of bio-coal containing iron ore composites under inert conditions and simulated blast furnace thermal profile. Composites with and without 10% bio-coal and sufficient amount of coke breeze to keep the C/O molar ratio equal to one were mixed and Portland cement was used as a binder. The self-reduction of composites was investigated by thermogravimetric analyses under inert atmosphere. To explore the reduction progress in each type of composite vertical tube furnace tests were conducted in nitrogen atmosphere up to temperatures selected based on thermogravimetric results. Bio-coal properties as fixed carbon, volatile matter content and ash composition influence the reduction of iron oxide. The reduction of the bio-coal containing composites begins at about 500 °C, a lower temperature compared to that for the composite with coke as only carbon source. The hematite was successfully reduced to metallic iron at 850 °C by using bio-coal, whereas with coke as a reducing agent temperature up to 1100 °C was required.

Keywords: devolatilization; torrefied biomass; bio-coal; volatile matter; reduction; blast furnace

---

1. Introduction

Ore-based ironmaking via the blast furnace (BF) dominates the metal supply for steel making [1]. For every ton of steel produced, on average 1.83 tons of CO_2 was emitted in 2017. According to the World Steel Association, the iron and steel industry accounts for approximately 7% to 9% of total world CO_2 emission [2]. Coke and coal as main reducing agents in the BF are the main contributors to CO_2 emitted during iron and steel making. The European Union (EU) has set a target to cut 80% of the CO_2 of fossil carbon, by 2050 [3]. In several studies the possible decrease in fossil CO_2 emissions by using biomass is reported [4–6].

The use of raw biomass as a reducing agent in the BF is difficult due to high moisture content, low content of fixed carbon (C_{fix}) as well as a high content of volatile matter (VM) and oxygen [7,8]. Different pretreatment technologies like pyrolysis [9], torrefaction [10], etc. can convert biomass into products with properties suitable for metallurgical applications. Pretreated biomass (bio-coals) has higher content of C_{fix}, lower contents of VM and oxygen, higher calorific value and can be pulverized without the formation of high ratio of fibers, properties which overall correspond more to the ones of injection coal [10,11]. It was reported that 70% of mass yield is retained as a solid product (char) during torrefaction process while 30% of mass is converted to gases [12]. The opposite is found for highly pretreated biomass, which is characterized by low char yield (~38% mass yield) as a big part of

VM was removed during process [13]. It can be an advantage to use torrefied product as reducing agent in composites containing iron oxide if remaining volatile contribute to the reduction. To select suitable bio-coals for use in composites with iron oxide an improved knowledge on the effect from bio-coal properties is required.

Swedish industry aims to reduce the CO_2 emission by different means and in the short term, this includes, e.g., improving the energy efficiency of the process, replacing fossil coal with reactive carbonaceous material like bio-coal (pretreated biomass) and in the longer term to use hydrogen. The use of biomass resources is a possible alternative in Sweden, as there are biomass resources available from forestland, areas estimated to be 28.1 million hectares [14]. The use of bio-coal as part of top charged briquettes also containing iron oxide has the potential to lower the thermal reserve zone temperature (TRZ) of the BF and thereby give a high replacement ratio to coke due to improved gas efficiency [15].

Partial replacement of coke by bio-coals and raw biomass in self-reducing mixtures or composites is reported in the literature [16–27]. The effect of the C/O molar ratio on the reduction behavior of iron ore has been studied earlier [16,24,25]. Najmi et al. [16] found that the reduction of iron oxide was enhanced when using pyrolized bio-coal in a composite tested isothermally at 1550 °C when compared to composites containing only coke. Liu et al. [24] studied the reduction behavior of bio-coal containing iron oxide composite pellets. The reduction was done isothermally at 1000–1300 °C in samples with different C/O molar ratio, it was found that as the C/O ratio increased the reduction rate also increased. Zuo et al. [25] compared the reduction behavior of iron oxide by using bio-coal, coal and coke. Self-reducing mixtures were reduced in a non-isothermal test procedure having a constant heating rate up to 1200 °C. It was found that bio-coal enhanced the reduction of hematite compared to coal and coke. Ubando et al. [27] found enhanced reduction using torrefied biomass in comparison to graphite in samples containing hematite to carbonaceous material in ratios of 2:1 and 1:1 heated non-isothermally up to 1200 °C.

Some of the recent research concerns bio-coals produced from similar original raw biomass pretreated at different temperatures and thereby getting different VM and ash content but having similar ash composition [19,23]. In isothermal reduction of iron ore containing composite in inert atmosphere at temperatures in the range of 800–1000 °C Hirokazu et al. [23] found that bio-coal having higher content of VM will enhance the reduction of iron ore more compared to bio-coal with lower VM content and the reduction rate of iron oxide was higher when bio-coal containing 18% VM was used in agglomerates compared to when coke was used. Ueda et al. [19] compared the reduction of composites containing carbon in CO/CO_2 and in Ar atmosphere up to 1200 °C but the impact from VM was not evaluated. It was stated that bio-coals, in general, enhance the reduction in comparison to coke.

Except for the work by Hirokazu et al. [23], there is no study on the influences of properties (C_{fix}, VM and ash composition) for different types of pre-treated bio-coal on the reduction of iron ore. In the current study, different pre-treated biomasses with different properties, e.g., in terms of VM and ash content as well as ash composition was used as a reducing agent in iron ore containing composite prepared as briquettes, with the aim to understand their impact on the reduction of iron ore. VM present in bio-coals contained in the composites should preferably contribute to the reduction and not being released and lost with the BF top gas. Composites containing only coke breeze are used as reference.

2. Materials and Methods

2.1. Material and Characterization

Composites were produced from iron oxide, coke breeze and four different bio-coals with Portland cement as a binder. Carbonaceous materials used are stated in Table 1, their proximate and ultimate analysis as analyzed by ALS Scandinavia AB using standard methods are presented in Table 2 and the composition of other raw materials can be seen in Table 3. The ash composition in bio-coal materials was reported in detail in a previous study [28], c.f. Table 4. As can be seen TSD has highest content of VM but contains ash with the lowest content of basic oxides, whereas TFR has almost as high VM content but contains ash with substantial amounts of basic oxides. HTT and CC has ash with similar contents of CaO but in HTT is balance by the SiO_2 content.

Table 1. Selected carbonaceous materials with abbreviations and their pretreatment temperatures and time.

Carbonaceous Materials	Temperature, $^\circ$C	Time, min	Abbreviation
Torrefied forest residue	286	6	TFR
Torrefied sawdust	297	6	TSD
High temperature torrefied	350	8	HTT
Charcoal	550	-	CC
Coke breeze	Up to 1100	-	CB

Table 2. Proximate and ultimate analysis results for carbonaceous materials (dry base).

Carbonaceous Materials	Proximate Analysis (wt %)			Ultimate Analysis (wt %)					Net Calorific Value (MJ/Kg)
	C_{fix}	VM	Ash	C_{tot}	H	N	S	O	
TFR	23.6	73.2	3.2	52.0	5.9	0.57	0.035	35.2	21.4
TSD	24.0	75.6	0.45	57.1	5.9	0.12	0.004	36.4	21.6
HTT	60.8	38.2	1.0	75.3	4.9	0.10	0.008	18.8	28.3
CC	80.7	18.6	0.70	87.0	3.4	0.25	<0.004	8.30	32.6
CB	88.2	0.7	11.1	87.3	<0.1	1.23	0.127	<0.10	29.7

C_{Fix} Fixed carbon; VM volatile matter; C_{tot} Total carbon; H hydrogen; N nitrogen; S sulphur; O oxygen.

Table 3. Chemical composition of pellet fines and cement, (wt %).

Sample	Fe	S	SiO_2	Al_2O_3	CaO	MgO	Na_2O	K_2O	MnO	V_2O_5
Pellet fines	66.7	0.0028	1.92	0.34	0.42	1.35	0.04	0.02	0.06	0.23
Cement	2.46	1.37	20.1	3.47	65.0	2.11	0.17	0.93	0.04	0.02

Table 4. Contents of metal oxides in the bio-coal materials (wt %, dry basis) [28].

Bio-Coals	Al_2O_3	CaO	SiO_2	Fe_2O_3	K_2O	MgO	MnO	Na_2O	P_2O_5	TiO_2
TFR	0.047	0.872	0.618	0.044	0.238	0.124	0.062	0.014	0.151	0.004
TSD	0.003	0.122	0.033	0.011	0.051	0.046	0.013	0.003	0.008	0.000
HTT	0.024	0.310	0.305	0.079	0.145	0.061	0.040	0.023	0.032	0.001
CC	0.006	0.317	0.028	0.009	<0.002	0.112	0.044	<0.009	0.006	0.001

The blends from which the briquettes were produced are stated in Table 5. Coke breeze used is screened off fines of metallurgical coke in the raw material handling at the Swedish steel producer SSAB Europe in Luleå [29]. The torrefied materials (TSD, TFR and HTT) were received from BioEndev [30] and the commercial charcoal was from Vindelkol [31]. All carbon-bearing materials were pulverized then sieved and the fraction 75–150 μm was used in the composites. The temperature for biomass pre-treatment affects the contents of VM, oxygen and C_{fix} (fixed carbon). Biomass torrefied at low temperature (TFR and TSD) has high contents of VM and oxygen but low content of C_{fix}; the opposite is the case for biomass pre-treated at high temperature (HTT and CC). The added iron oxide is in the form of olivine pellet fines screened off at LKAB, and these consist mainly of hematite. The particle size distribution for iron ore was analyzed to d_{90} ~163 μm by laser diffraction-based particle size analyzer (CILAS 1064, Micromeritics Instrument Corporation, Orléans, France). Cement was used as it is a common binder in industrially produced briquettes.

Table 5. Recipes of self-reducing composites.

Composite	Composite Composition, (wt %)				
	Pellet Fines	Coke Breeze	Bio- Coal	Cement	Abbreviation
FR composite	68.1	13.8	10.0	8.0	FRC
SD composite	68.1	13.8	10.0	8.0	SDC
HTT composite	71.5	10.5	10.0	8.0	HTTC
CC composite	73.3	8.67	10.0	8.0	CCC
Bio-coal free composite	74.0	18.0	-	8.0	BFC

2.2. Composite Preparation

The composites were prepared by mixing pulverized carbonaceous materials, iron ore and cement according to the recipes in Table 5. A total of 100 g of a homogenous mixture of each blend was prepared, water was added, and mixing continued for five minutes. Approximately 2.5 g of the blend was put in a cylindrical steel mold 10 mm in diameter and 14 mm in height. The blend was compacted for 2.5 min under a load of 1800 kg/cm^2. Produced briquettes were covered by plastic sheets to maintain humidity and cured at 30–40 °C for 7 days. Cured briquettes were dried overnight at 105 °C before conducting a reduction test.

The carbon in carbonaceous material to oxygen bound to iron in hematite pellet fines was prepared aiming for C_{fix}/O molar ratio equal to one. Considering the total carbon to oxygen bound to hematite, the ratio will be larger due to the content of carbon in VM present in bio-coals. Chemical composition of composites shown in Table 6 was determined by Degerfors Laboratorium AB in Sweden using a Thermo ARL 9900 X-ray fluorescence (XRF) instrument (Thermo Fisher Scientific Inc, Waltham, MA, USA) operating a rhodium tube at 50 kV 50 mA. The sample was pressed in cellulose briquettes before analysis. Carbon and sulfur content for each composite was determined by using LECO CS-444 instrument (LECO Corporation, Lakeview Ave, St. Joseph, MI, USA). The calculated VM content in Table 6 is based on contents in each bio-coal and briquette recipes. FRC and SDC have a higher content of VM compared to other composites.

Table 6. Chemical composition of self-reducing composites, (wt %).

Composite	C_{tot}	CaO	MgO	SiO_2	Al_2O_3	Na_2O	K_2O	P_2O_5	S	Fe_2O_3	Cl	VM*
FRC	16.8	9.24	1.97	7.68	2.08	0.22	0.44	0.06	0.038	59.7	0.01	7.4
SDC	15.6	10.0	1.85	7.87	2.31	0.19	0.32	0.04	0.036	60.0	<0.01	7.7
HTTC	16.5	8.97	1.86	7.05	2.19	0.14	0.54	0.05	0.034	60.8	-	3.9
CCC	15.2	8.41	1.92	7.10	2.09	0.26	0.52	-	0.030	62.9	0.02	1.9
BFC	15.2	7.11	2.39	9.51	3.48	0.29	0.27	0.06	0.048	59.5	<0.01	0.10

VM*: VM calculated from the recipe for each composite.

2.3. Thermogravimetric Analysis

Reduction tests were conducted in the thermogravimetric analyzer, Netzsch STA 409, (sensitivity ±1 μg) attached to a Quadrupole Mass Spectrometer (QMS, Netzsch, Selb, Germany) to monitor the mass-loss and off-gas analysis, respectively. As reducing agents, the bio-coals specified in Tables 1 and 2 were used. The thermogravimetric analyzer used in this study is described in detail earlier [32]. The heating rate during thermogravimetric analyses (TGA) was chosen to simulate a possible BF thermal profile as given in Figure 1 but under inert conditions with argon gas (99.999% purity) at a rate of 200 mL/min. The thermal profile and heating rates are 20 °C/min in the temperature range of 0–500 °C, 4 °C/min between 500–850 °C, 1 °C/min between 850–950 °C and 3 °C/min between 950–1100 °C. Finally, the sample was kept 1 h at 1100 °C before cooling with 20 °C/min to room temperature. To eliminate error from the buoyancy effect, correction measurements were carried out and each test was repeated three times to ensure the accuracy of results and the results were consistent without any significant variation. The samples were cooled down to room temperature at a rate of 20 °C/min and kept dry in a desiccator for subsequent characterization.

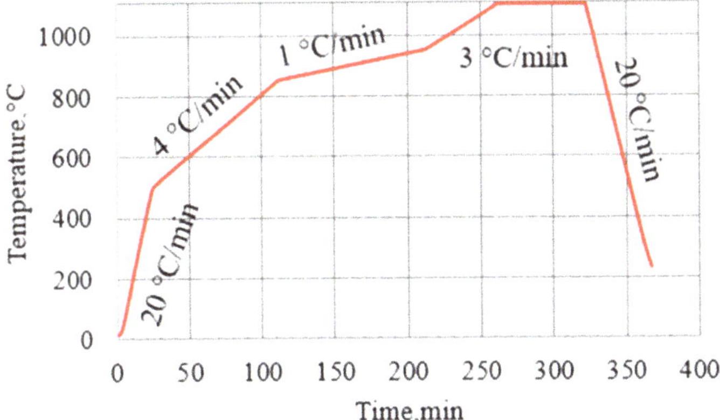

Figure 1. Simulated BF thermal profile used in TGA and for tube furnace tests.

2.4. Interrupted Reduction Tests of Composites

Interrupted tests were carried out in N_2 gas atmosphere (purity 99.996% and a flow rate of 5 L/min) introduced from the top in a tube furnace with a schematic layout as shown in Figure 2. Platinum/Platinum–Rhodium Type S thermocouples were used for measurement and control of temperature when heating the tube furnace with Super Kanthal heating elements. The sample was placed in an alumina crucible hanging in the hot zone of the 120 cm long alumina tube with an inner diameter of 8.65 cm. The sample was heated according to the thermal profile stated in Figure 1 to a predetermined temperature (500, 680, 740, 850, 950 °C). When the desired temperature was reached, the sample was transferred into the cooling chamber and the flow rate of N_2 gas was increased to 10 L/min and 5 L/min from the top and into the cooling chamber, respectively. The samples were kept in a desiccator until characterization was carried out.

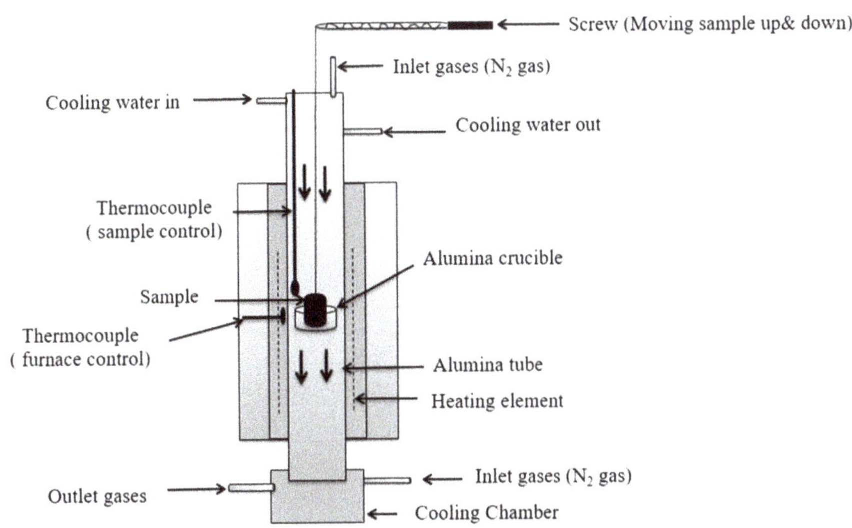

Figure 2. Schematic layout of the tube furnace.

2.5. Characterization

2.5.1. Mineralogy

X-ray diffraction (XRD) was used to determine the phases present in reduced samples. All samples were ground in a mortar and analyzed using a Panalytical Empyrean X-ray diffractometer equipped with copper Kα radiation of 45 kV and 40 mA (Malvern Panalytical, Almeo, Netherlands). Diffraction patterns were measured in a 2θ range of 10° to 90° during 15 min for each sample.

2.5.2. Morphology

The textures of composites were investigated on polished samples mounted in epoxy using light optical microscope (LOM, Nikon ECLIPSE E600 POL, Minato/Tokyo, Japan) and scanning electron microscope Zeiss Gemini Merlin (FE-SEM, Carl Zeiss AG, Oberkochen, Germany) equipped with energy-dispersive X-ray spectroscopy (EDS) unit for elemental analysis. Prior to SEM investigation, the polished surface of mounted samples was coated with a thin layer of tungsten. For SEM, a beam operation voltage of 20 kV and current of 1.0 nA was used.

3. Results

3.1. TGA/QMS and DTG

Figure 3 shows the TGA/QMS curves for bio-coal containing composites (BCC) and bio-coal free composites (BFC) using simulated BF thermal profile. The TG analysis shows that BCC has higher mass loss compared to BFC at lower temperatures and that the mass loss can be divided into three distinct regions, see Figure 3. The mass loss of FRC and SDC are quite similar, slightly higher for FRC and both are higher than for other composites as presented in Table 7. DTG (the first derivative of mass loss) analysis showed that FRC has a higher mass loss rate than SDC at ~300 °C as seen in the Region I in Figure 4. By increasing temperature up to 850 °C, the mass loss of FRC, SDC and HTTC increased and are higher than for CCC and BFC. At temperatures above 850 °C, the mass loss rate is high for CCC as seen in Figure 4. The mass-loss rate for all composites increased as seen in Region III (up to 1100 °C) while the main mass loss for CCC and BFC lies at 850–950 °C and 950–1100 °C, respectively.

Table 7. Mass loss (wt %) of composites during TGA tests.

Composites	Region I	Region II	Region III		
	Up to 500 °C	Up to 850 °C	850–950 °C	950–1100 °C	Isothermal Part at 1100 °C
FRC	7.41	18.4	6.59	14.1	0.30
SDC	6.83	19.0	6.36	13.3	0.29
HTTC	3.99	16.8	10.7	10.1	0.49
CCC	1.83	14.2	19.2	2.67	0.17
BFC	0.92	5.31	8.99	20.4	0.68

The QMS analysis shows ionized hydrocarbons with one, two or four carbon atoms per molecule and gases (CO, CO_2, and H_2) in the off-gas during the reduction test. However, the lengths of carbon chains in the released hydrocarbon during bio-coal devolatilization were probably initially longer before thermal decomposition and excitation in the QMS. The release of hydrocarbons in the low-temperature region occurred at the lowest temperature for FRC and SDC as seen in Figure 3 with the highest mass loss rate for FRC as seen in Figure 4.

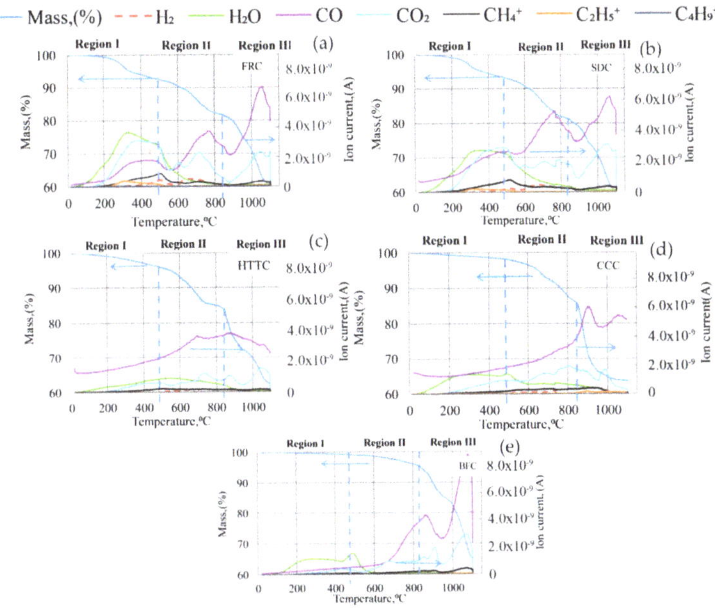

Figure 3. TGA-QMS analysis of self- reducing composites (**a**) FRC; (**b**) SDC; (**c**) HTTC; (**d**) CCC and (**e**) BFC in Argon up to 1100 °C using BF simulated thermal profile.

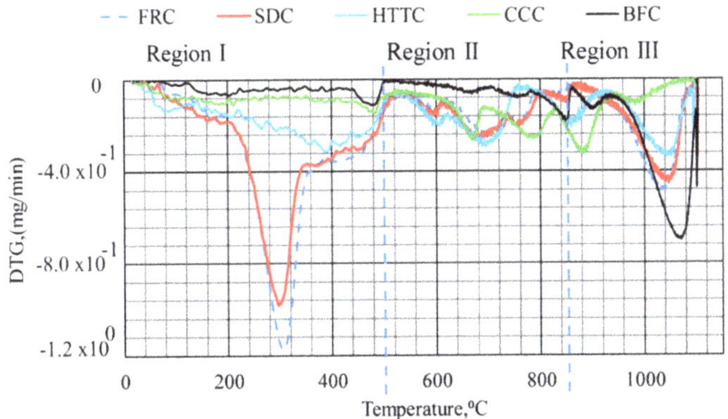

Figure 4. DTG curves of self- reducing composites during TGA tests in Ar gas up to 1100 °C using simulated BF thermal profile.

3.2. Interrupted Reduction Tests of Composites

The interrupted tests were conducted to determine the progress of reduction with different carbonaceous materials present in each composite. The temperatures for interruption of tests were selected based on the difference in mass loss curves seen from TGA in the three regions pointed out in Figure 3. Results on mass loss in interrupted tests conducted in a tube furnace are presented in Table 8.

Table 8. Mass loss (wt %) of composites during interrupted tests.

Temperature, °C	FRC	SDC	HTTC	CCC	BFC
500	7.39	6.55	3.48	1.55	0.54
680	10.3	11.2	7.21	4.12	1.81
740	14.3	15.6	13.0	7.07	1.83
850	17.3	18.1	16.1	10.2	4.15
950	19.6	20.1	21.2	19.8	6.62

3.3. XRD Analysis

XRD analysis conducted for composites collected after interrupted tests in the tube furnace and finalized tests in TGA is shown in Figure 5. The following is observed;

- Fe_2O_3 was detected in the BFC sample from the test interrupted at 500 °C.
- Fe_3O_4 was detected in all bio-coal containing samples from tests interrupted at 500 °C, for BFC it was first detected at 740 °C.
- *FeO* was detected in the samples from tests interrupted at 680 °C for SDC, at 740 °C for FRC and HTTC, at 850 °C in CCC and at 950 °C for BFC.
- *Fe* was detected in samples from tests interrupted at 850 °C for SDC and HTTC, at 950 °C for FRC and CCC, at 1100 °C for BFC.

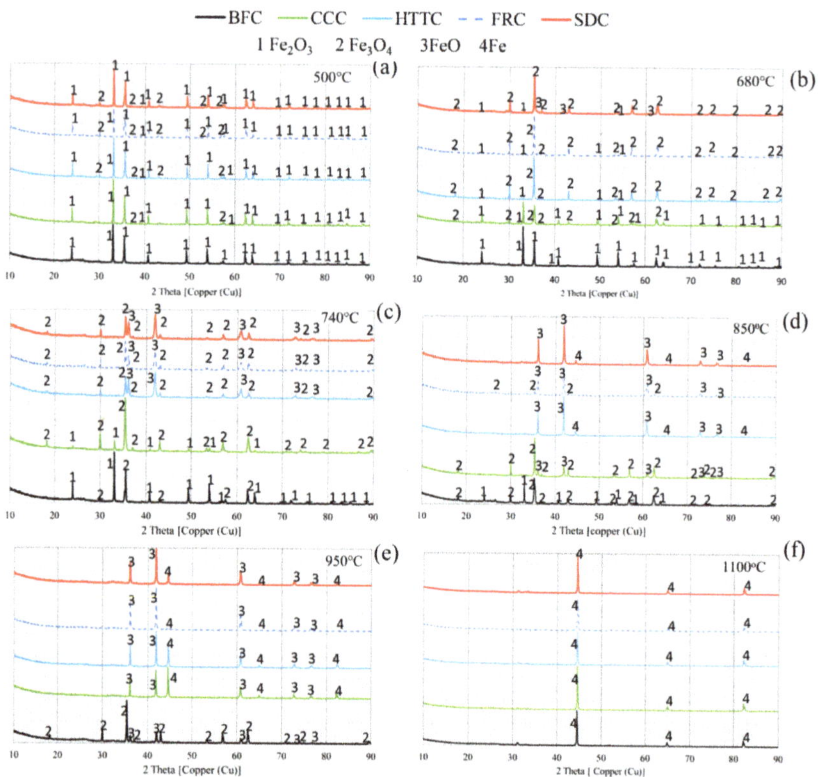

Figure 5. XRD analysis of reduced composite treated in a tube furnace at (**a**) 500 °C; (**b**) 680 °C; (**c**) 740 °C; (**d**) 850 °C; (**e**) 950 °C and (**f**) final sample up to 1100 °C in TGA.

3.4. Evaluation of Composite Structure during Reduction

The structures of composites from interrupted tests were examined in LOM and SEM. Figure 6 presents the microstructures of two different types of composites collected after tests interrupted at 680 °C. The conversion of hematite to magnetite occurred initially at the outer layer of hematite particles in all bio-coal containing composites which can be seen for FRC in Figure 6a. In composites collected after tests interrupted at ~740 °C, the transformation of hematite to magnetite had proceeded further in SDC and a porous structure and cracks had been formed as seen in Figure 7a while the transformation of hematite to magnetite was still in the outer layer in HTTC as seen in Figure 7b. The duct structure of bio-coal was seen in FRC and quite clear in CCC as seen in Figures 6a and 6b, respectively.

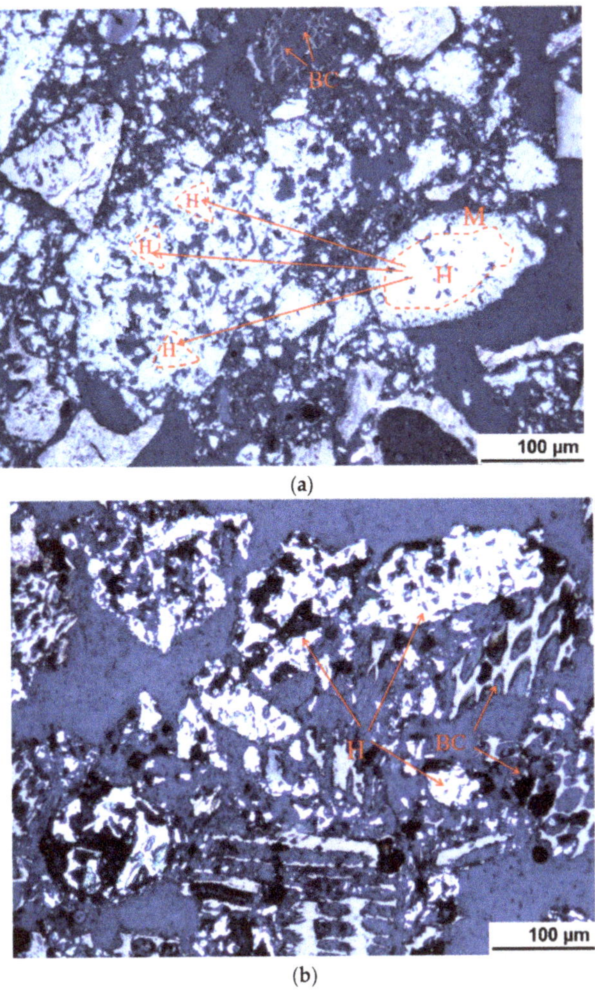

Figure 6. Typical textures observed in LOM (magnification 20x) in composites from tests interrupted at 680 °C. (**a**) FRC and (**b**) CCC. H = hematite (white-grey), M = magnetite (grey) and BC = bio-coal.

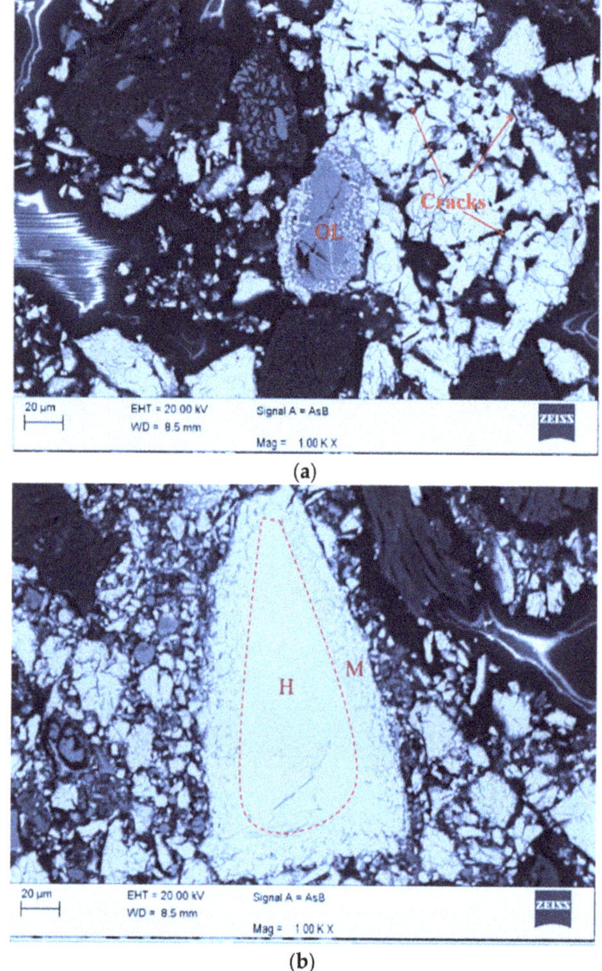

(a)

(b)

Figure 7. SEM image (magnification 1000×), of composites from tests, interrupted at 740 °C. (**a**) SDC, (**b**) HTTC, H = hematite, M = magnetite, and Ol = Olivine.

4. Discussion

The effect of different pre-treated biomass on the self-reduction of agglomerates containing iron oxide and coke breeze has been investigated. Composites were analyzed in TGA and treated in vertical tube furnace up to pre-defined temperatures. To evaluate the impact of bio-coal on reduction, samples from interrupted tests were investigated by XRD, LOM and SEM. The mass loss occurring in TGA is assumed to depend on partial devolatilization of VM ($biomass \rightarrow Volatile\,(gases + tar) + bio - char$), gasification $\left(C_{fix} + CO_2 \leftrightarrow 2CO\right)$, and reduction of iron oxides ($Fe_2O3 \rightarrow Fe3O4 \rightarrow FeO \rightarrow Fe$). CO formed during devolatilization contributed to the reduction of iron oxides.

The TGA results and interrupted test up to 500 °C showed that FRC has higher mass loss than all other composites, and a higher mass-loss rate was clear from DTG analysis at ~300 °C. Although TSD and TFR have quite similar content of VM, the higher mass loss detected at low temperature for FRC could be due to the presence of higher contents of catalyzing components such as CaO and K_2O in TFR than in TSD. This may affect the release of H_2O during reaction as the ion current for H_2O is highest for FRC although that the H-content is similar as for SDC. The impact of ash composition on the devolatilization behavior of bio-coals was indicated in a previous study [28] where bio-coal with a higher content of catalyzing components released the VM at lower temperature, and this may restrict its contribution to the reduction. High intensity of H_2O followed by CO_2 and H_2 detected in QMS analyses for FRC indicates that some of the CO is consumed in the reaction. It is possible that this affects the contribution to reduction of iron oxide in FRC. From XRD analysis, it was seen that FeO had been formed in samples interrupted at 740 °C for FRC while it was detected in samples interrupted at 680 °C for SDC.

QMS analysis showed that the intensity of CO gas in regions I and II is highest for SDC, reaching a maximum at ~760 °C, as shown in Figure 8. Part of this CO gas might originate from thermal decomposition of VM as well as from the Boudouard reaction in which CO_2 reacts with remaining carbon above 700 °C as reported by Butterman [33]. Furthermore, it is observed that FeO is formed in SDC at lower temperatures than for other composites and SDC has the highest mass loss rate in region II (up to 850 °C).

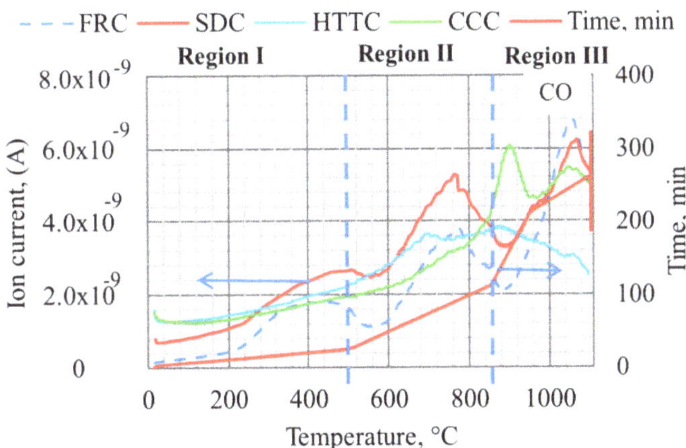

Figure 8. Off-gas analysis of CO for different types of bio-coal containing composites (BCC).

The progress of the reduction is indicated in LOM by more cracks in SDC samples interrupted at 740 °C while it was detected only in the outer layer for HTTC at the same temperature. Wustite was not detected in CCC samples interrupted below 850 °C. CO generation from CCC samples occurred mainly above 850 °C, likely due to C_{fix} starting to react, c.f. Figure 8, simultaneously with a higher mass loss rate for CCC. This can be compared with BFC in which wustite and iron were first detected by XRD in samples interrupted at 950 °C and in the final sample at 1100 °C, respectively. The higher intensity for CO seen in QMS results for BFC at temperatures above 950 °C is likely due to the transformation step from FeO to Fe_{met} and solution loss reactions taking place above 950 °C.

XRD analysis shows also that magnetite is formed in all BCC samples at 500 °C, but not in BFC composites. CO formed in Region I in all BCC, but not in BFC, may have contributed to the earlier start of reduction in BCC. It is shown that FeO is formed at a lower temperature in the composites containing more VM, indicating either that more reactive char is formed or that volatiles still existing above 500 °C, as in SDC, also contributes to the reduction. It is also shown that BFC containing coke breeze demand higher temperature for gasification reaction and reduction compared to BCC, in agreement with Ueda et al. [19].

The results show that if bio-coal with high volatile content is selected for use in a composite, a low content of catalyzing components is preferable as volatiles are released at higher temperatures and thereby could contribute to the reduction.

The investigated composites are potential innovative materials to be added to the BF together with coke. By energy consuming reactions, the thermal reserve zone temperature is reduced and the overall reduction efficiency of the BF improved. When selecting carbon source for the composites, the results show that coke breeze demands higher temperature for reaction compared to bio-coals. TSD could be the most suitable material to be utilized in composites in the BF. This material has higher content of volatile matter content with a lower content of catalyzing components, which means that the volatiles can be released at higher temperatures and thereby can contribute to the reduction. Additionally, outside the scope of this paper, it is known that the yield when producing bio-coal is high for products as TSD.

5. Conclusions

The reduction behavior of composites containing hematite and different bio-coals was studied and compared with bio-coal free composite (coke as carbon source). The following conclusion can be drawn:

- Bio-coal present in an agglomerate enhances the reduction of hematite, in comparison with coke, and the reduction starts at a lower temperature.
- Bio-coal with a high content of volatiles decreases the temperature for the formation of wustite more than bio-coal with lower content of volatiles.
- The presence of ash components catalyzing the volatilization to occur at lower temperatures results in a higher temperature for the formation of wustite.

Author Contributions: A.A.E.-T. conceived and designed the experiments, performed the experiments, analyzed the data, wrote the first draft of the paper, reviewed and wrote the paper; L.S.Ö., B.B. and H.M.A. supervised, reviewed and contributed to the writing of the paper. All authors have read and agreed to the published version of the manuscript.

Funding: The research was funded by the Bio-agglomerate project, grant number 156334.

Acknowledgments: Financial support from the Swedish Energy Agency (Energimyndigheten) for research within the bio-agglomerate project is gratefully acknowledged. For additional support, the following companies and institutions are acknowledged: BioEndev, Swerim AB, SSAB Merox, LKAB and SSAB special steel. The research has been partly financed by CAMM, Center of Advanced Mining and Metallurgy at LTU.

Conflicts of Interest: The authors declare no conflict of interest.

References

1. Geerdes, M.; Chaigneau, R.; Kurunov, I. *Modern Blast Furnace Ironmaking: An. Introduction (2015)*, 3rd ed.; Ios Press: Amsterdam, The Netherlands; Holland, The Netherlands, 2015; pp. 1–215.
2. World Steel Association. *Steel's Contribution to a Low Carbon Future and Climate Resilient Societies–Worldsteel Position Paper*; World Steel Association: Brussels, Belgium, 2017; Available online: https://www.worldsteel.org/publications/bookshop/product-details~{|}Steel-s-Contribution-to-a-Low-Carbon-Future--2018-update-~{|}PRODUCT~{|}Steel-s-Contribution-to-a-Low-Carbon-Future~{|}.html (accessed on 14 April 2018).
3. Wang, C.; Mellin, P.; Lövgren, J.; Nilsson, L.; Yang, W.; Salman, H.; Hultgren, A.; Larsson, M. Biomass as blast furnace injectant–Considering availability, pretreatment and deployment in the Swedish steel industry. *Energy Convers. Manag.* **2015**, *102*, 217–226. [CrossRef]
4. Feliciano-Bruzual, C. Charcoal injection in blast furnaces (Bio-PCI): CO_2 reduction potential and economic prospects. *J. Mater. Res. Technol.* **2014**, *3*, 233–243. [CrossRef]
5. Babich, A.; Senk, D.; Fernandez, M. Charcoal behaviour by its injection into the modern blast furnace. *ISIJ Int.* **2010**, *50*, 81–88. [CrossRef]
6. Mandova, H.; Leduc, S.; Wang, C.; Wetterlund, E.; Patrizio, P.; Gale, W.; Kraxner, F. Possibilities for CO_2 emission reduction using biomass in European integrated steel plants. *Biomass Bioenergy* **2018**, *115*, 231–243. [CrossRef]

7. Fick, G.; Mirgaux, O.; Neau, P.; Patisson, F. Using biomass for pig iron production: A technical, environmental and economical assessment. *Waste Biomass Valoris* **2014**, *5*, 43–55. [CrossRef]

8. Chen, W.; Du, S.; Tsai, C.; Wang, Z. Torrefied biomasses in a drop tube furnace to evaluate their utility in blast furnaces. *Bioresour. Technol.* **2012**, *111*, 433–438. [CrossRef]

9. Bridgwater, A. The production of biofuels and renewable chemicals by fast pyrolysis of biomass. *Int. J. Global Energy Issues* **2007**, *27*, 160–203. [CrossRef]

10. Van der Stelt, M.J.C.; Gerhauser, H.; Kiel, J.H.A.; Ptasinski, K.J. Biomass upgrading by torrefaction for the production of biofuels: A review. *Biomass Bioenergy* **2011**, *35*, 3748–3762. [CrossRef]

11. Shankar Tumuluru, J.; Sokhansanj, S.; Hess, J.R.; Wright, C.T.; Boardman, R.D. REVIEW: A review on biomass torrefaction process and product properties for energy applications. *Ind. Biotechnol.* **2011**, *7*, 384–401. [CrossRef]

12. Bergman, P.C.A.; Boersma, A.R.; Zwart, R.W.R.; Kiel, J.H.A. *Torrefaction for Biomass Co-Firing in Existing Coal-Fired Power Stations "BIOCOAL"*; Technical Report; ECN: Petten, The Netherlands; Holland, The Netherlands, 2005; ECN-C–05-013.

13. Hanif, M.U.; Capareda, S.C.; Kongkasawan, J.; Iqbal, H.; Arazo, R.O.; Baig, M.A. Correction: Effects of pyrolysis temperature on product yields and energy recovery from Co-feeding of cotton gin trash, cow manure, and microalgae: A simulation study. *PLoS ONE* **2016**, *11*, e0156565. [CrossRef]

14. Official Statistics of Sweden. *SKOGSDATA 2018*; Technical Report; The Swedish University of Agricultural Sciences: Umeå, Sweden, May 2018; Available online: https://www.slu.se/en/Collaborative-Centres-and-Projects/the-swedish-national-forest-inventory/forest-statistics/skogsdata/ (accessed on 12 December 2018).

15. Mousa, E.; Lundgren, M.; Sundqvist Ökvist, L.; From, L.; Robles, A.; Hällsten, S.; Sundelin, B.; Friberg, H.; El-Tawil, A. Reduced carbon consumption and CO_2 emission at the blast furnace by use of briquettes containing torrefied sawdust. *J. Sustain. Metall.* **2019**, *5*, 391–401. [CrossRef]

16. Najmi, N.H.; Yunos, M.; Diyana, N.F.; Othman, N.K.; Asri Idris, M. Characterisation of reduction of iron ore with carbonaceous materials. *Solid State Phenom.* **2018**, *280*, 433–439. [CrossRef]

17. Machado, J.; Osorio, E.; Vilela, A.; Babich, A.; Senk, D.; Gudenau, H. Reactivity and conversion behaviour of Brazilian and imported coals, charcoal and blends in view of their injection into blast furnaces. *Steel Res. Int.* **2010**, *81*, 9–16. [CrossRef]

18. Ng, K.W.; Giroux, L.; MacPhee, T.; Todoschuk, T. Combustibility of charcoal for direct injection in blast furnace ironmaking. *Iron Steel Technol.* **2012**, *9*, 70.

19. Ueda, S.; Watanabe, K.; Yanagiya, K.; Inoue, R.; Ariyama, T. Improvement of reactivity of carbon iron ore composite with biomass char for blast furnace. *ISIJ Int.* **2009**, *49*, 1505–1512. [CrossRef]

20. Strezov, V. Iron ore reduction using sawdust: Experimental analysis and kinetic modelling. *Renew. Energy* **2006**, *31*, 1892–1905. [CrossRef]

21. Srivastava, U.; Kawatra, S.K.; Eisele, T.C. Production of pig iron by utilizing biomass as a reducing agent. *Int. J. Miner. Process.* **2013**, *119*, 51–57. [CrossRef]

22. Guo, D.; Hu, M.; Pu, C.; Xiao, B.; Hu, Z.; Liu, S.; Wang, X.; Zhu, X. Kinetics and mechanisms of direct reduction of iron ore-biomass composite pellets with hydrogen gas. *Int. J. Hydrogen Energy* **2015**, *40*, 4733–4740. [CrossRef]

23. Konishi, H.; Ichikawa, K.; Usui, T. Effect of residual volatile matter on reduction of iron oxide in semi-charcoal composite pellets. *ISIJ Int.* **2010**, *50*, 386–389. [CrossRef]

24. Liu, Z.; Bi, X.; Gao, Z.; Liu, W. Carbothermal reduction of iron ore in its concentrate-agricultural waste pellets. *Adv. Mater. Sci. Eng.* **2018**, *2018*, 1. [CrossRef]

25. Zuo, H.; Hu, Z.; Zhang, J.; Li, J.; Liu, Z. Direct reduction of iron ore by biomass char. *Int. J. Miner. Metall. Mater.* **2013**, *20*, 514. [CrossRef]

26. Guo, D.; Li, Y.; Cui, B.; Chen, Z.; Luo, S.; Xiao, B.; Zhu, H.; Hu, M. Direct reduction of iron ore/biomass composite pellets using simulated biomass-derived syngas: Experimental analysis and kinetic modelling. *Chem. Eng. J.* **2017**, *327*, 822. [CrossRef]

27. Ubando, A.T.; Chen, W.; Ong, H.C. Iron oxide reduction by graphite and torrefied biomass analyzed by TG-FTIR for mitigating CO_2 emissions. *Energy* **2019**, *180*, 968. [CrossRef]

28. El-Tawil, A.A.; Ökvist, L.S.M.; Ahmed, H.; Björkman, B. Devolatilization kinetics of different types of bio-coals using thermogravimetric analysis. *Metals* **2019**, *9*, 168. [CrossRef]

29. Lundgren, M. Development of Coke Properties during the Descent in the Blast Furnace. Ph.D. Thesis, Luleå University of Technology, Luleå, Sweden, 2013.
30. Bioendev. Technology. Available online: http://www.bioendev.se/technology/torrefaction/ (accessed on 19 September 2019).
31. Vindelkol. Technology. Available online: http://www.vindelkol.se/ (accessed on 19 September 2019).
32. Ahmed, H.M.; Persson, A.; Okvist, L.S.; Bjorkman, B. Reduction behaviour of self-reducing blends of in-plant fines in inert atmosphere. *ISIJ Int.* **2015**, *55*, 2082–2089. [CrossRef]
33. Butterman, H.C.; Castaldi, M.J. CO_2 as a carbon neutral fuel source via enhanced biomass gasification. *Environ. Sci. Technol.* **2009**, *43*, 9030–9037. [CrossRef] [PubMed]

Article

Mechanism of CaF$_2$ under Vacuum Carbothermal Conditions for Recovering Nickel, Iron, and Magnesium from Garnierite

Qiang Wang [1,2,3,4], Xupeng Gu [1,2,3,4], Tao Qu [1,2,3,4,*], Lei Shi [1,2,3,4], Mingyang Luo [1,2,3,4], Bin Yang [1,2,3,4] and Yongnian Dai [1,2,3,4]

[1] State Key Laboratory of Complex Non-ferrous Metal Resources Clear Utilization, Kunming University of Science and Technology, Kunming 650093, China; wqiang_doc@163.com (Q.W.); xvpenggu@163.com (X.G.); ahdyshilei@163.com (L.S.); 18712711093@163.com (M.L.); kgyb2005@126.com (B.Y.); daiyn@cae.cn (Y.D.)

[2] National Engineering Laboratory for Vacuum Metallurgy, Kunming University of Science and Technology, Kunming 650093, China

[3] Key Laboratory of Vacuum Metallurgy for Nonferrous Metal of Yunnan Province, Kunming University of Science and Technology, Kunming 650093, China

[4] Faculty of Metallurgical and Energy Engineering, Kunming University of Science and Technology, Kunming 650093, China

* Correspondence: qutao_82@126.com

Received: 15 November 2019; Accepted: 9 January 2020; Published: 15 January 2020

Abstract: Nickel laterite ore is divided into three layers and the garnierite examined in this study belongs to the third layer. Garnierite is characterized by high magnesium and silicon contents. The main contents of garnierite are silicates, and nickel, iron, and magnesium exist in silicates in the form of lattice exchange. Silicate minerals are difficult to destroy so are suitable for smelting using high-temperature pyrometallurgy. To solve the problem of the large amounts of slag produced and the inability to recycle the magnesium in the traditional pyrometallurgical process, we propose a vacuum carbothermal reduction and magnetic separation process to recover nickel, iron, and magnesium from garnierite, and the behavior of the additive CaF$_2$ in the reduction process was investigated. Experiments were conducted under pressures ranging from 10 to 50 Pa with different proportions of CaF$_2$ at different temperatures. The experimental data were obtained by various methods, such as thermogravimetry, differential scanning calorimetry, scanning electron microscopy, energy dispersive spectrometry, X-ray diffraction, and inductively coupled plasma atomic emission spectroscopy. The analysis results indicate that CaF$_2$ directly reacted with Mg$_2$SiO$_4$, MgSiO$_3$, Ni$_2$SiO$_4$, and Fe$_2$SiO$_4$, which were isolated from the bearing minerals, to produce low-melting-point compounds (FeF$_2$, MgF$_2$, NiF$_2$, etc.) at 1315 and 1400 K. This promoted the conversion of the raw materials from a solid–solid reaction to a liquid–liquid reaction, accelerating the mass transfer and the heat transfer of Fe–Ni particles, and formed Si–Ni–Fe alloy particles with diameters of approximately of 20 mm. The smelting materials appeared stratified, hindering the reduction of magnesium. The results of the experiments indicate that at 1723 K, the molar ratio of ore/C was 1:1.2, the addition of CaF$_2$ was 3%, the recovery of Fe and Ni reached 82.97% and 98.21% in the vacuum carbothermal reduction–magnetic separation process, respectively, and the enrichment ratios of Fe and Ni were maximized, reaching 3.18 and 9.35, respectively.

Keywords: garnierite; vacuum carbothermal reduction; mechanism; CaF$_2$; recovery

1. Introduction

Nickel is an important metal in the modern world, used in stainless steel, electroplating, rechargeable batteries, and super alloys [1,2] to improve peoples' quality of life, and widely used in improving human health in nickel-containing medical devices, medical artificial bones, stents, and new anticancer drugs [3–6]. With the increase of demand for nickel and the rapid depletion of nickel sulfide ore, more attention is being paid to low-grade nickel laterite ore [7–9]. Generally, nickel laterite ore can be divided into three categories: limonitic ore, transition ore, and garnierite [10–12]. According to the different compositions of every layer, nickel laterite ores are treated using different methods, including pyrometallurgical and hydrometallurgical methods [13]. Hydrometallurgical techniques are more applicable to limonitic ore, which includes ammonia–ammonium carbonate leaching, pressure acid leaching (PAL), and high-pressure acid leaching (HPAL) [14–20]. Garnierite is more suitable for pyrometallurgical methods due to the high content of magnesium. At present, the mature industrial pyrometallurgical methods include rotary kilns-electric furnaces (RKEF) and Krupp–Ren [21–23].

Previous studies reported that CaF_2 is used as additive in pyrometallurgy of nickel laterite ore. Ma et al. [24] studied the influence of CaF_2 in the nickel laterite ore reduction process. The results indicated that the aggregation of nickel and iron particles increased significantly from 1 to 200 µm after adding CaF_2, which indicates that CaF_2 can effectively reduce the surface tension of the newly formed alloy. This is beneficial for magnetic separation process and increases the nickel grade. Cao et al. [25] determined the production and enrichment rates of nickel and iron under different smelting conditions by changing the amount of added reducing agent and comparing the different production rates of ferronickel with CaF_2, CaF_2, and hydrated lime. However, the authors did not report the optimal conditions, nor did they specifically explore the mechanism of CaF_2 and hydrated lime in the reduction process.

Previous studies only focused on the extraction of Ni and Fe, and ignored the extraction of higher values magnesium. The behavior of CaF_2 in the reduction process has not been reported in the literature. Therefore, we propose a vacuum carbothermal reduction–magnetic separation process to recover nickel, iron, and magnesium from garnierite, and investigated the behavior of the added CaF_2 in the reduction process. In this process, magnesium was reduced to magnesium vapor and condensed to obtain magnesium metal [26,27]. The experimental flow chart is shown in Figure 1.

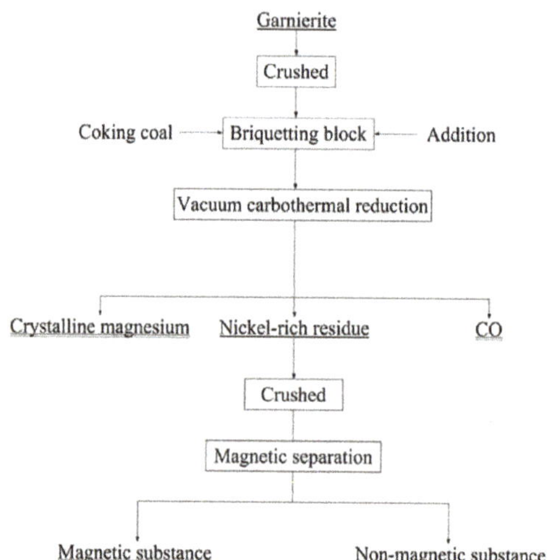

Figure 1. Work flow of the experimental procedure.

2. Materials and Methods

2.1. Raw Materials

The garnierite sample used in this study was obtained from Yuanjiang, Yunnan province, China. The X-ray diffraction (XRD) analysis and the chemical analysis of the sample are shown in Figure 2 and Table 1. According to the XRD analysis, we found that the major phases of garnierite are lizardite, nepouite, kaolinite, and quartz. CaF_2 was used as an additive in this study. The compositions of CaF_2 and the coking coal are listed in Tables 2 and 3, respectively.

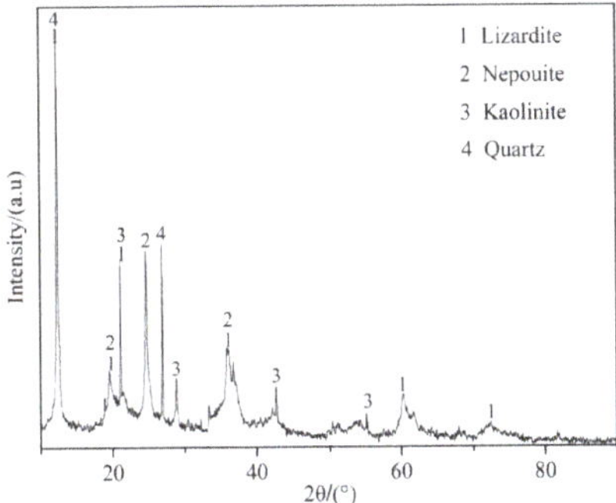

Figure 2. X-ray diffraction (XRD) patterns of garnierite.

Table 1. The composition of the garnierite (wt. %).

Component	SiO$_2$	MgO	Fe	Ni	Co	Al$_2$O$_3$	Others
Content	38.82	22.83	12.66	0.72	0.03	4.57	20.37

Table 2. The composition of the CaF$_2$ (wt. %).

Component	CaF$_2$	Si	Fe	Heavy Metal	Chloride	Sulfate	Nitride
Content	≥98.5	≤0.01	≤0.003	≤0.003	≤0.01	≤0.05	≤0.005

Table 3. The composition of the coking coal (wt %).

Component	C	Ash	Moisture	Volatile	K (g)
Content	85.83	12.02	0.14	2.01	7187.2

2.2. Experimental Method and Equipment

All garnierite and coking coal were crushed and screened to less than 74 μm, and the ore and C were mixed in a molar ratio of 1:1.2. The masses of CaF_2, which were 0%, 3%, 6%, 9%, and 12%, were studied with regard to their behavior and effect on the experimental results. In the first, the mixed materials were formed into pellets of Φ20 mm × 20 mm at 5 MPa and then placed in a corundum crucible. Heating was applied under pressures ranging from 10 to 50 Pa by placing the corundum crucible into a graphite crucible in a vacuum furnace. The system temperature was raised from room temperature to the desired temperature and was held this temperature for 120 min. After, the reaction product was cooled to room temperature and removed. The schematic model of the vacuum distillation furnace for experiment is shown in Figure 3. For the nickel-rich residue, we used a vibrating grinding machine to crush the material to approximately 74 μm, and we placed the crushed materials into a planetary ball mill for 4 h of ball milling, allowing the materials to be mixed thoroughly and uniformly. After the ball milling was completed, the materials were removed, sampled, and sent for analyses, and magnetic separation was conducted for the remaining residue.

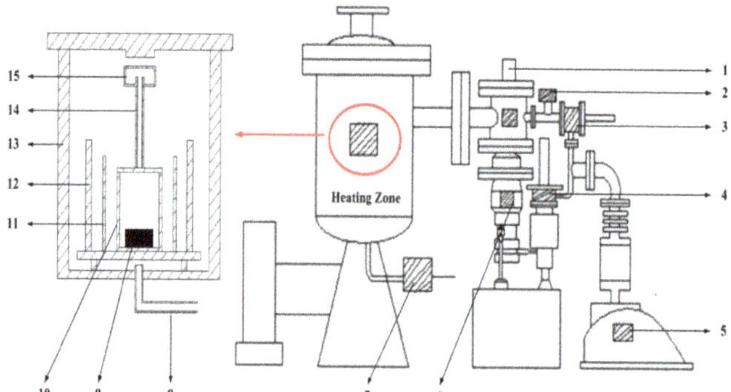

Figure 3. Schematic model of the vacuum distillation furnace: 1, high vacuum valve; 2, release valve; 3, bypass valve; 4, main valve; 5, rotary vane pump; 6, diffusion pump; 7, inflation valve; 8, vacuum pipe; 9, raw materials; 10, graphite crucible; 11, graphite heater; 12, heat shield; 13, water cooled walls; 14, graphite condensing pipe; and 15, condensation.

The remaining nickel-rich residue was ground and screened to 45 μm, and then wet magnetic separation was conducted. The N52 trademark of neodymium iron, with a surface magnetic field strength approximately 1.44–1.48 T, was used to adsorb the magnetic material. A magnetic substance was obtained after three magnetic separations.

The recovery α_1 of the reduction process is expressed as:

$$\alpha_1 = M_a/M_0, \tag{1}$$

where M_a and M_0 refer to the masses of metals in the nickel-rich residue and in the ore, respectively. The recovery α_2 of the reduction magnetic separation process is expressed as:

$$\alpha_2 = M_b/M_a, \tag{2}$$

where M_a and M_b refer to the masses of metals in the magnetic substances and in the nickel-rich residue, respectively. The recovery α of the whole process is expressed as:

$$\alpha = \alpha_1 \times \alpha_2. \tag{3}$$

2.3. Analysis Methods

The nickel-rich residue was identified with X-ray diffraction (XRD) Dmax-R diffractometer (Rigaku, Tokyo, Japan, Cu Kα radiation, 40 mA, 50 kV). The diffraction angle (2θ) was scanned from 10° to 90° in 4° increments. The morphology and the elemental composition of the particular regions were determined using scanning electron microscopy (SEM, TM-3030 Plus, HITACHI, Tokyo, Japan) equipped with energy-dispersive X-ray spectroscopy (EDS, INCA, Oxford, UK). Chemical analysis of the nickel-rich residue was performed using inductively coupled plasma atomic emission spectroscopy (ICP-OES) with an Optima 8000 (Perkin Elmer, Waltham MA, USA).

3. Results and Discussion

3.1. Experimental Phenomena and XRD/EDS Analysis

During the removal of the Fe–Ni-rich residue, we found different phenomena at the end of the experiment; no melting occurred at the experimental temperature when the addition of CaF_2 was 0%. When the addition of CaF_2 was 3%, the raw materials appeared to experience micro-melting at 1623 and 1723 K, however, the raw materials melted completely at 1823 K, as shown in Figure 4.

Figure 4. The reduced slag at (**a**) 1623 K, (**b**) 1723 K, and (**c**) 1823 K with 3% CaF$_2$.

The left part of Figure 5 is the picture of the material removed after the vacuum carbothermal reduction. The black fluffy material in the picture is located on the top of the corundum crucible. After pouring the black fluffy material, we found that the material in the corundum crucible completely melted, with a layer of gray white material solidified with no hole on the surface on the top of the melted material. After the corundum crucible was broken, we found many large cavities in the melted material, and the color of solidified material below was gray black. Finally, large metal particles were found at the bottom of the corundum crucible. According to the above phenomena, we drew a layering diagram of the materials in the corundum crucible, as shown in the right part of Figure 5.

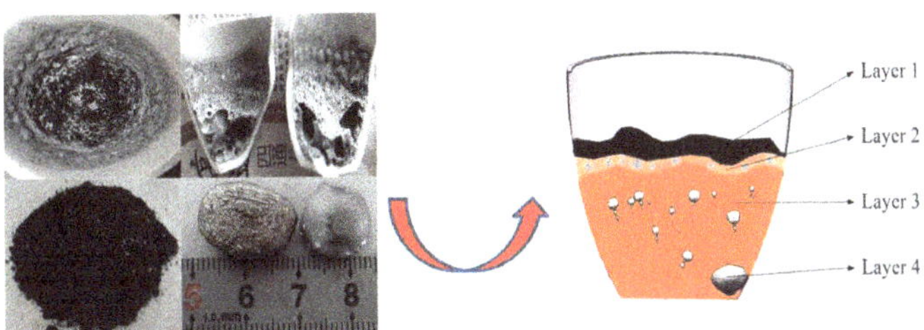

Figure 5. Phenomena of stratification for the residue after reduction at the experimental temperature with more than 3% CaF$_2$.

XRD and SEM/EDS analysis were conducted to determine the phase of black fluffy material and the composition of silver-white metal particles of the first layer, as shown in Figure 6. According to XRD analysis, the black fluffy material was mainly carbon, and the ferrosilicon should be the silver-white metal particles, which mixed with the black matter. SEM/EDS analysis showed that the carbon content reached 75.28%, the remaining materials were mainly Fe and F, and the contents of Fe and F were 12.71% and 5.30%, respectively.

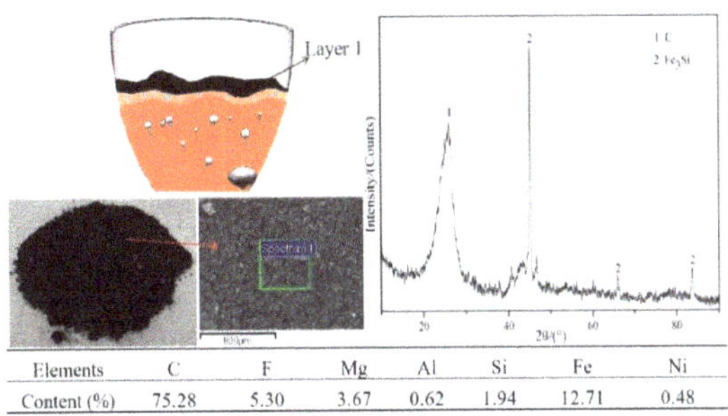

Elements	C	F	Mg	Al	Si	Fe	Ni
Content (%)	75.28	5.30	3.67	0.62	1.94	12.71	0.48

Figure 6. XRD pattern and energy-dispersive X-ray spectroscopy (EDS) analysis of Layer 1.

The thermodynamic analysis of Equation (4) performed at 50 Pa showed that the initial reaction temperature of the production of FeF_2 was 1398 K, and FeF_2 will directly sublimate at 1373 K. We considered that the source of F was the volatilization of FeF_2, with part of the Fe sourced from the ferrosilicon mixed in black matter, and the other part from the volatilization of FeF_2. The equation can be expressed as:

$$Fe_2SiO_{4(s)} + 6C_{(s)} + CaF_{2(s)} = FeSi_{(s)} + CaC_{2(s)} + FeF_{2\,(g)} + 4CO_{\,(g)}$$
$$\Delta G_T = 1731.126 - 1.238T \text{ kJ/mol.} \tag{4}$$

XRD and SEM/EDS analysis were performed to determine the phase and composition of the second layer, as shown in Figure 7. The XRD results showed that the white matter phase of the second layer was $MgAl_2O_4$, and the SEM/EDS results indicated that the white matter mainly contained Mg, Al, and O, at 14.28%, 26.58%, and 50.38%, respectively.

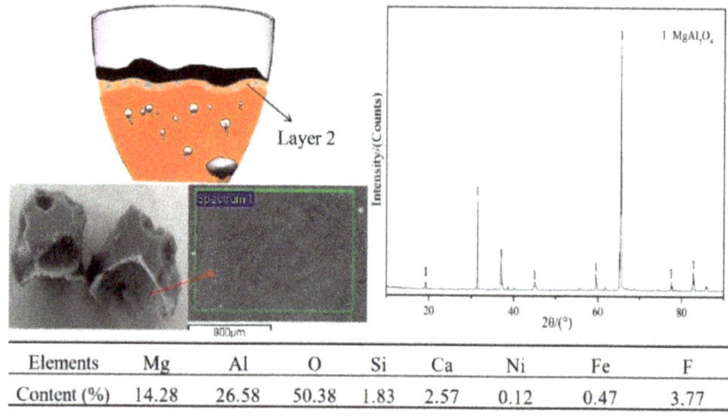

Elements	Mg	Al	O	Si	Ca	Ni	Fe	F
Content (%)	14.28	26.58	50.38	1.83	2.57	0.12	0.47	3.77

Figure 7. XRD pattern and EDS analysis of Layer 2.

Combined with the thermodynamic calculation, Equations (5) and (6) showed that the reaction initial temperature for the formation of $MgAl_2O_4$ was 875 K, so the reaction to form $MgAl_2O_4$ could occur at the experimental conditions. The reason of Al_2O_3 did not reduce by carbon was the initial temperature of the reduction reaching 1691 K under the pressure of 50 Pa, which is much larger than the reaction initial temperature of $MgAl_2O_4$. The $MgAl_2O_4$ is denser than carbon, so the carbon floats above the $MgAl_2O_4$. The equations can be expressed as:

$$Mg_2SiO_{4(s)} + 2Al_2O_{3(s)} = 2MgAl_2O_{4(s)} + SiO_{2(s)}$$
$$\Delta G_T = 19.255 - 0.022T \text{ kJ/mol,} \tag{5}$$

$$Al_2O_{3(s)} + 3C_{(s)} = 2Al_{(s)} + 3CO_{\,(g)}$$
$$\Delta G_T = 1344.640 - 0.795T \text{ kJ/mol.} \tag{6}$$

XRD and SEM/EDS analysis were performed to determine the phase and composition of the material in the third layer as well, as shown in Figure 8. XRD analysis revealed that the main phases in the third layer were Mg_2SiO_4 and $MgAl_2O_4$; SEM/EDS analysis showed that the main elements of this layer were Mg and Si, with contents of 45.97% and 29.10%, respectively, mixed with a small amount of F, Fe, Ca, and Al. According to the thermodynamic calculation in Equation (7), the reaction initial temperature of Mg_2SiO_4 under the pressure of 50 Pa was 1472 K, which is higher than the melt initial temperature. According to the analysis of the first and second layers, we found that after the melting, the carbon and the materials stratified, so most of the magnesium was not reduced. The equation can be expressed as:

$$3Mg_2SiO_{4(s)} + 6C_{(s)} + 2CaF_{2(s)} = 6Mg_{(g)} + 6CO_{(g)} + 2CaSiO_{3(s)} + SiF_{4(g)} \tag{7}$$
$$\Delta G_T = 4297.884 - 2.920T \ kJ/mol.$$

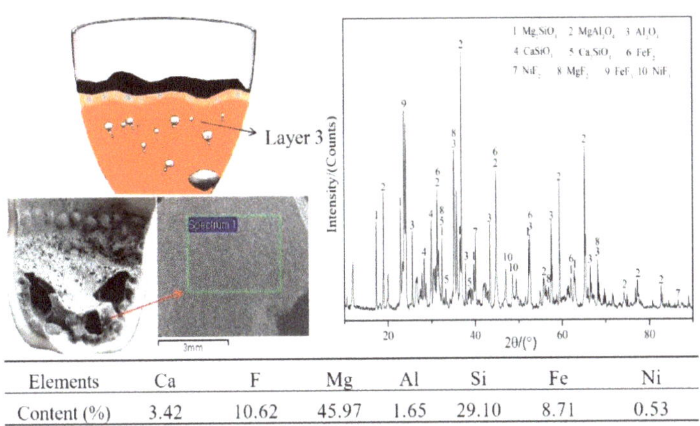

Elements	Ca	F	Mg	Al	Si	Fe	Ni
Content (%)	3.42	10.62	45.97	1.65	29.10	8.71	0.53

Figure 8. XRD pattern and EDS analysis of Layer 3.

According to the experimental findings, due to the block created by second layer, many bubbles were trapped underneath, so the Mg vapor could not overflow. This further hindered the reduction, so a large amount of Mg_2SiO_4 remained in the materials, and a large amount of Mg_2SiO_4 formed after the end of the experiment. The density of Mg_2SiO_4 is greater than that of $MgAl_2O_4$, resulting in $MgAl_2O_4$ floating above the Mg_2SiO_4.

XRD and SEM/EDS analysis were performed to determine the phase and composition of the fourth layer, as shown in Figure 9. XRD analysis revealed that the fourth layer was mainly Fe_XSi_Y in various compositions. SEM/EDS analysis revealed that the layer mainly elements were Fe, Ni, and Si, with contents of 66.90%, 4.15%, and 25.72%, respectively.

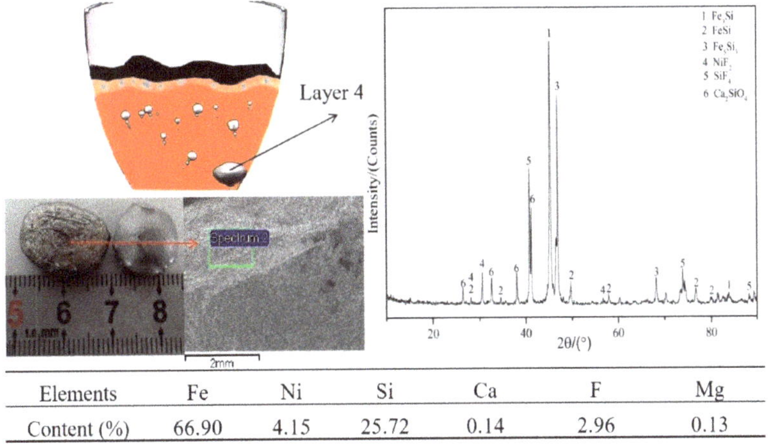

Elements	Fe	Ni	Si	Ca	F	Mg
Content (%)	66.90	4.15	25.72	0.14	2.96	0.13

Figure 9. XRD pattern and EDS analysis of Layer 4.

From the thermodynamic calculations of Equations (4) and (8)–(10), we found that the formation of Fe, Fe_XSi_Y, Ni, and FeNi formed before the materials melted. Under the pressure of 50 Pa, the initial temperature of reactions were 1344, 808, 618, and 1389 K, respectively. After the materials melted, the mass transfer and heat transfer of Ni and Fe accelerated, forming a large particle of Fe–Ni–Si ternary alloy. Due to being denser than Mg_2SiO_4, it sunk into the bottom of the crucible to form the fourth layer. The equations can be expressed as:

$$3Fe_2SiO_{4(s)} + 6C_{(s)} + 2CaF_{2(s)} = 6Fe_{(s)} + 6CO_{(g)} + 2CaSiO_{3(s)} + SiF_{4\,(g)}$$
$$\Delta G_T = 1336.421 - 1.653T \text{ kJ/mol,} \tag{8}$$

$$3Ni_2SiO_{4(s)} + 6C_{(s)} + 2CaF_{2(s)} = 6Ni_{(s)} + 6CO_{(g)} + 2CaSiO_{3(s)} + SiF_{4\,(g)}$$
$$\Delta G_T = 1110.829 - 1.797T \text{ kJ/mol,} \tag{9}$$

$$Ni_2SiO_{4(s)} + 6C_{(s)} + CaF_{2(s)} = NiSi_{(s)} + CaC_{2(s)} + NiF_{2(s)} + 4CO_{(g)}$$
$$\Delta G_T = 1384.267 - 1.055T \text{ kJ/mol.} \tag{10}$$

At the end of the experiment, a dense condensate with a silvery white metallic luster was found on the lid, as shown in Figure 10. XRD and SEM/EDS analysis were performed to determine the phase and composition of the condensate. XRD analysis revealed that condensate was mainly calcium–silicon compounds. SEM/EDS analysis showed that the elemental contents of Si, Ca, and Fe were 54.85%, 16.58%, and 8.47%, respectively. From the thermodynamic calculations of Equations (7)–(9) and (11)–(13), we found that the initial temperatures of the formation of SiF_4 and SiO were 1472, 808, 618, 1085, 961, and 1431 K under the pressure of 50 Pa, respectively. Therefore, the Si in the condensate was mainly produced from the condensation of SiF_4 and SiO, the Fe from the sublimation of FeF_2, and Ca from the condensation of calcium compounds. The equations can be expressed as:

$$Fe_2SiO_{4(s)} + 3C_{(s)} = SiO_{(g)} + 2Fe_{(s)} + 3CO_{(g)}$$
$$\Delta G_T = 1044.890 - 0.962T \text{ kJ/mol,} \tag{11}$$

$$Ni_2SiO_{4(s)} + 3C_{(s)} = SiO_{(g)} + 2Ni_{(s)} + 3CO_{(g)}$$
$$\Delta G_T = 968.751 - 1.008T \text{ kJ/mol,} \tag{12}$$

$$SiO_{2(s)} + C_{(s)} = SiO_{(g)} + CO_{(g)}$$
$$\Delta G_T = 697.671 - 0.487T \text{ kJ/mol.} \tag{13}$$

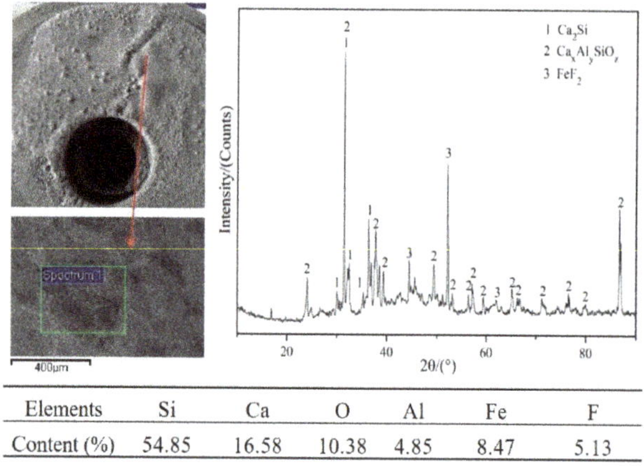

Elements	Si	Ca	O	Al	Fe	F
Content (%)	54.85	16.58	10.38	4.85	8.47	5.13

Figure 10. XRD pattern and EDS analysis of the condensate on the graphite crucible lid.

According to the above analysis of the phase and composition of each layer of materials, we found that the first layer was mainly coking coal, the small silver-white metal particles were Fe–Ni–Si, and the main component of the second transition layer was $MgAl_2O_4$. The third layer was mainly Mg_2SiO_4 with a small amount of $MgAl_2O_4$ and calcium–silicon compounds, and the large metal particles in the fourth layer were Fe–Ni–Si ternary alloy. The condensate on the lid was mainly Si.

3.2. Behavior of CaF$_2$

The thermogravimetry-differential scanning calorimetry (TG and DSC) curves of the garnierite are shown in Figure 11. We observed three peaks at different temperatures. According to a previous report [28], the first endothermic peak at 352 K represents the loss of the adsorption water, the second endothermic peak at 885.4 K represents the dehydroxylation of the lizardite, and the third strong exothermic peak at 1097.8 K represents the decomposition of $Mg_3Si_2O_7$. The process is expressed in Equations (14) and (15).

$$Mg_3Si_2O_5(OH)_{4(s)} = Mg_3Si_2O_{7(s)} + 2H_2O_{(g)} \tag{14}$$

$$Mg_3Si_2O_{7(s)} = Mg_2SiO_{4(s)} + MgSiO_{3(s)} \tag{15}$$

The TG and DSC curves of the garnierite with CaF$_2$ are shown in Figure 12. By comparing Figures 11 and 12, we found two more endothermic peaks at 1315 and 1400 K. The two endothermic peaks represented the conversion of the raw materials from solid to liquid, which produced low-melting-point compounds such as FeF$_2$, NiF$_2$, and MgF$_2$. The thermodynamic calculations in Equations (4), (10) and (16) were performed at 50 Pa. The equations can be expressed as:

$$Mg_2SiO_{4(s)} + SiF_{4\,(g)} = 2MgF_{2(s)} + 2SiO_{2(s)}$$
$$\Delta G_T = -271.244 + 0.221T \text{ kJ/mol.} \tag{16}$$

Based on Equations (4) and (10), the initial temperatures of the two reactions were 1398 and 1312 K, which corresponded to the two endothermic peaks in Figure 12.

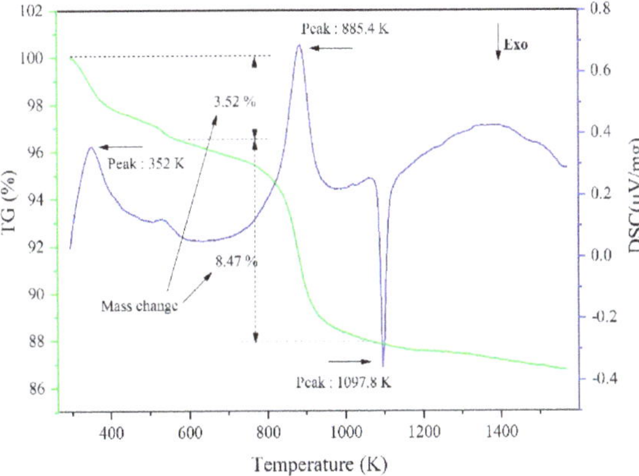

Figure 11. Thermogravimetry-differential scanning calorimetry (TG and DSC) curves of the raw materials.

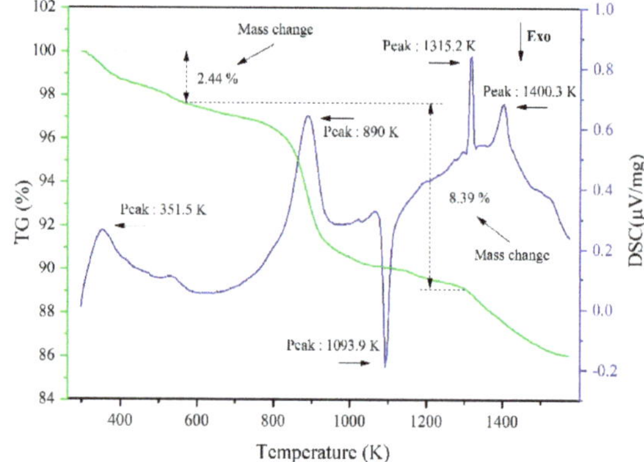

Figure 12. TG and DSC curves of the raw materials in presence of CaF$_2$.

3.3. Analysis of the Reduction Process

After the vacuum carbothermal reduction, chemical analysis was performed on the Fe–Ni-rich residue. According to the calculation, we obtained the direct recovery of Fe and Ni, which is plotted in Figures 13 and 14, respectively. The figures indicate that with the increase in CaF$_2$, the grades and direct recovery of Fe and Ni first rose and then fell, with maximal grades and direct recovery achieved at 3% CaF$_2$. With increasing temperature, the curves in the figures divided into two parts. The addition of CaF$_2$ at less than or equal to 3% was the first part, in which the grades and direct recovery of Fe and Ni first rose and then fell, with the best grades and direct recovery achieved at 1723 K. The addition of CaF$_2$ at more than 3% was the second part, in which the grades and direct recovery of Fe and Ni showed a downward trend.

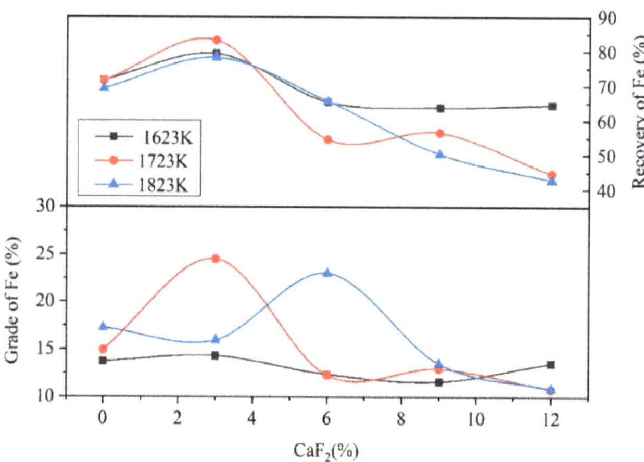

Figure 13. The effect of additives and temperature on Fe enrichment (%), the upper three lines in the figure represents the recovery of Fe, and the lower three lines in the figure represents the grade of Fe.

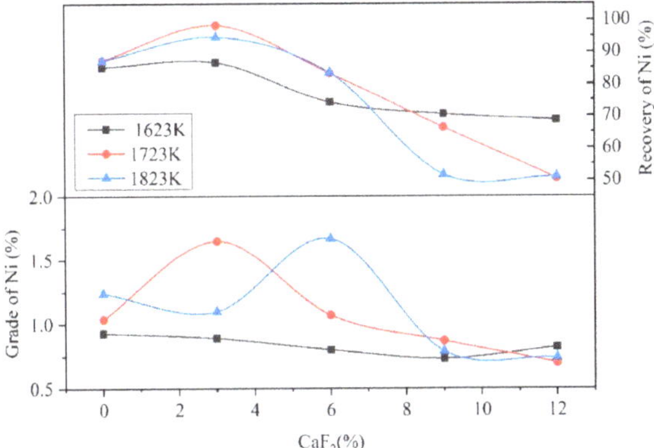

Figure 14. The effect of additives and temperature on Ni enrichment (%), the upper three lines in the figure represents the recovery of Ni, and the lower three lines in the figure represents the grade of Ni.

The reasons for these results are as follows:

(1) When the addition of CaF_2 was 0%, low-melting-point compounds did not form in the raw materials, and the reactions struggle to occur in the solid state in the reduction process, which caused the low grades and direct recovery of Fe and Ni.

(2) When the addition of CaF_2 was 3%, the CaF_2 could react with the raw materials to form low-melting-point compounds, such as FeF_2, NiF_2, and MgF_2. The reduction process transformed from a solid–solid reaction to a solid–liquid reaction, which accelerated the heat transfer and the mass transfer of the materials. The Fe and Ni aggregated and grew easily in the molten materials and formed small visible Fe–Ni particles on the surface of the Fe–Ni-rich residue. The raw materials were not completely melted in the reduction process due to the low addition of CaF_2. Therefore, Fe and Ni were sufficiently reduced in the raw materials, which results in high grades and direct recovery of Fe and Ni.

(3) When the addition of CaF_2 was more than 3%, a large amount of the low-melting-point compounds formed in the raw materials, which caused the raw materials to change from a solid–solid reaction to a liquid–liquid reaction. This in turn caused the reduction of Fe and Ni to aggregate and grow into large metal particles with diameters of 20 mm. However, the coking coal would not melt at the experimental temperature due to its high melting point, and the density of the coking coal was low, which caused stratification of the molten materials and floating in the upper layer. This resulted in the reducing agent being unable to fully contact the raw materials, preventing the reduction process from occurring, which drastically reduced the grades and direct recovery of Fe and Ni.

(4) Due to excessively high temperatures during the reduction process, the raw materials melted completely at temperature of 1823 K with 3% CaF_2, hindering the reduction process, which reduced the grades and the direct recovery of Fe and Ni. Therefore, the grades and direct recovery of Fe and Ni were lower than 1723 K with 3% CaF_2.

Our analyses demonstrate that the addition of 3% CaF_2 at 1723 K is the optimal condition.

According to Figure 15, with increasing temperature and CaF_2, the grades and recovery of Si showed a downward trend because Si would volatilize in the form of SiO and SiF_4 at high temperature, which resulted in low grades and direct recovery of Si. The equations are expressed in Equations (7)–(9) and (11)–(13).

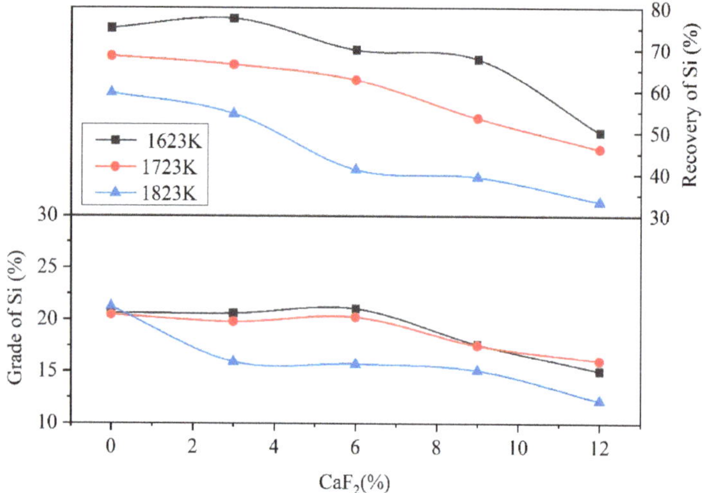

Figure 15. The effect of additives and temperature on Si enrichment (%). The upper three lines in the figure represents the recovery of Si, and the lower three lines in the figure represents the grade of Si.

According to Equations (7)–(9) and (11)–(13), the initial temperatures of these reactions were 1472, 808, 618, 1086, 961, and 1432 K, respectively. All the reactions could occur under experimental conditions. These analysis results were consistent with the above four explanations.

Figure 16 shows that the removal rate of Mg increased with increasing temperature and CaF_2 content. However, the removal rate was low because the raw materials changed from a solid–solid reaction to a liquid–liquid reaction with increasing CaF_2 content, which caused the reduction of Mg to only occur in the early period. However, after the materials melted completely, the coking coal and the raw materials were stratified, which preventing the coking coal from contacting the raw materials. The steam of the Mg could not pass through the transition layer, resulting in the removal rate of Mg increasing slowly. The best example of this was at a temperature of 1823 K with 3% CaF_2, where the removal rate of Mg was significantly lower than 1723 K with 3% CaF_2. These analysis results were consistent with the findings in Section 3.2.

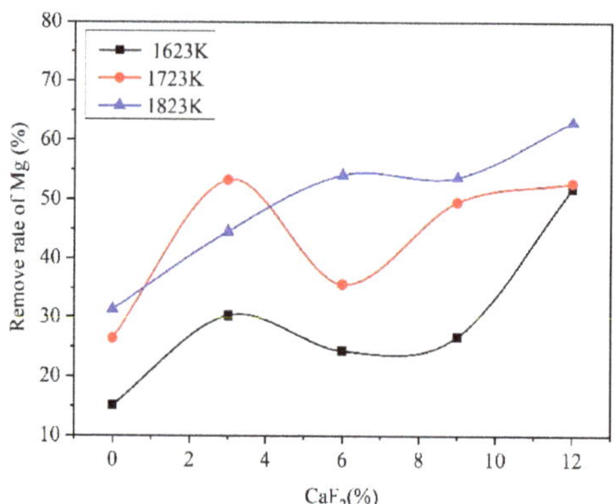

Figure 16. The effect of additives and temperature of Mg removal rate (%).

As seen in Figures 13–16, we can conclude that adding 3% CaF_2 is beneficial to the recovery of Ni and Fe, however, the recovery of Si and Mg are not good enough. At 1723 K, the recovery of Si is less than 45% and the recovery of Mg is less than 55%. The effect of temperature and the adding amount of CaF_2 on the recovery of Si and Mg showed the opposite trend, that is, with the increase of temperature and the amount of CaF_2, the recovery of Si showed a downward trend, while the recovery of Mg showed an upward trend. This become a contradiction point for recovering of garnierite comprehensively, therefore, the recovery of Si and Mg will not be ideal if only adding CaF_2. Our previous research [29] has proved that CaO can greatly enhance the recovery of Si and Mg. We could consider the mixture of CaF_2–CaO as an additive to achieve the purpose of comprehensive recovery of valuable metals from garnierite.

3.4. Phase Analysis of Fe–Ni-Rich Residue

Figures 17 and 18 show that the main phases of the Fe–Ni-rich residue were forsterite, spinel, and FeSi, as well as a small amount of $CaSiO_3$ and low-melting-point compounds (FeF_2, NiF_2, and MgF_2). According to the literature, forsterite, Fe_2SiO_4, and Ni_2SiO_4 are mainly produced after the dehydroxylation of lizardite. Forsterite can react with coking coal to form Mg steam at high temperatures. However, the Fe–Ni-rich residue was mainly forsterite in this experiment, which indicated that most of the forsterite did not participate in the reduction reaction.

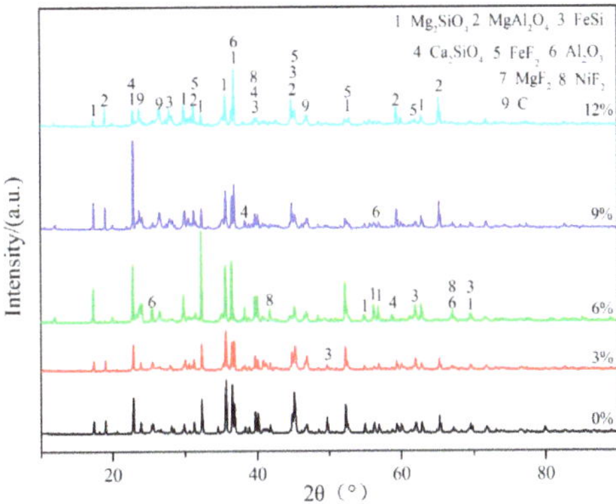

Figure 17. XRD patterns of the Fe–Ni-rich residue at 1723 K with 0–12% CaF_2.

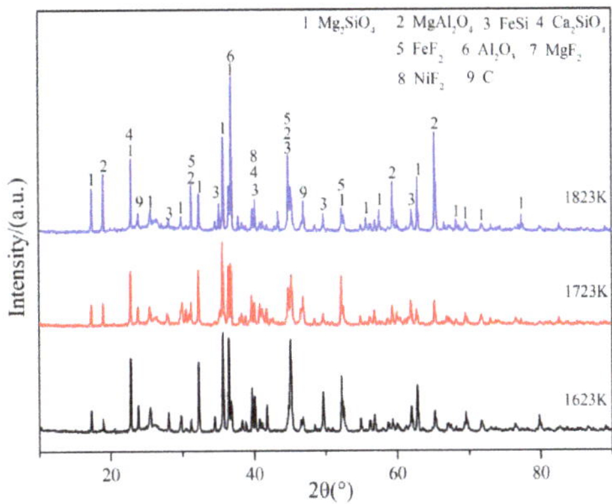

Figure 18. XRD patterns of the Fe–Ni-rich residue at different temperatures with 3% CaF_2.

After the lizardite decomposed, the CaF_2 reacted with Fe_2SiO_4 and Ni_2SiO_4 to form low-melting-point compounds (FeF_2, NiF_2, and MgF_2), as shown in Figure 17. With increases in CaF_2 and temperature, the raw materials changed from the solid to the liquid state. This caused some of the forsterite to react with the coking coal during the melting process, although a large part of forsterite could not react with coking coal due to the stratification after the raw materials melted completely, which caused a large amount of the forsterite to remain in the Fe–Ni-rich residue. The proportion of Ca in the raw materials was less than that of Si, the reduction of Fe and Ni could have combined with Si to form Si–Fe–Ni ternary alloy particles, and the proportion of Ni was low. Therefore, the Si–Fe–Ni ternary alloy showed up as an Fe–Si peak in the XRD pattern. The forsterite and the Al_2O_3 could have reacted to form spinel, which caused a spinel peak to be found in the Fe–Ni-rich residue. The equations were expressed as Equations (5) and (6) in Section 3.1.

According to Equations (5) and (6), the initial temperatures of the two reactions were 875 K and 1691 K. These results indicate that the reaction in Equation (5) more easily occurred than that in Equation (6), which meant the spinel could be produced under experimental conditions.

As shown in Figure 18, at 1723 K with 3% CaF_2, the forsterite peak was weakest and the Fe–Si peak was strongest, which indicates that the reduction of the raw materials was effective under these conditions.

3.5. Mechanism of CaF_2 in the Reduction Process

TG-DSC and thermodynamic analysis indicated that the solid materials began to melt at 1315 K and 1400 K and produced low melting point eutectic materials (NiF_2 and FeF_2) as heating progressed. In the reduction progress, the raw materials reacted with CaF_2 to produce SiF_4, and forsterite reacted with SiF_4 to form another low melting point eutectic material, MgF_2, which further accelerated the melting of the materials, causing the materials to transition from solid to liquid. The melting point of NiF_2 is 1653 K at 1 atm, the boiling point is 1772 K, the melting point of FeF_2 is 1243 K, the boiling point is 1373 K, and FeF_2 undergoes sublimation at 1373 K. At 50 Pa, the melting points of NiF_2 and FeF_2 decrease, therefore, once FeF_2 is formed, it undergoes sublimation and NiF_2 volatilizes simultaneously. According to the thermodynamics calculations, the initial temperature where MgF_2 forms is 1227 K, the melting point of MgF_2 is 1534 K, and the boiling point is 2533 K, therefore, MgF_2 would not have volatilized under the experimental conditions.

The 20-mm-diameter Fe–Ni–Si ternary alloy collected in the experiment showed that the melted materials accelerated the heat transfer and the mass transfer of Fe and Ni, promoting the FeNi particles to aggregate and grow. With increasing temperature and the increase in addition of CaF_2, the volatility of Si in the form of SiO and SiF_4 were promoted, causing a sharp drop in the recovery of Si. Due to the density of coking coal being less than that of $MgAl_2O_4$, $MgAl_2O_4$ being less dense than Mg_2SiO_4, and Mg_2SiO_4 being less dense than Fe–Ni–Si ternary alloy, the materials stratified. The stratification of the raw materials and coking coal hindered the reduction of Mg, causing a slow increase in the removal rate of Mg, as shown in Figure 19.

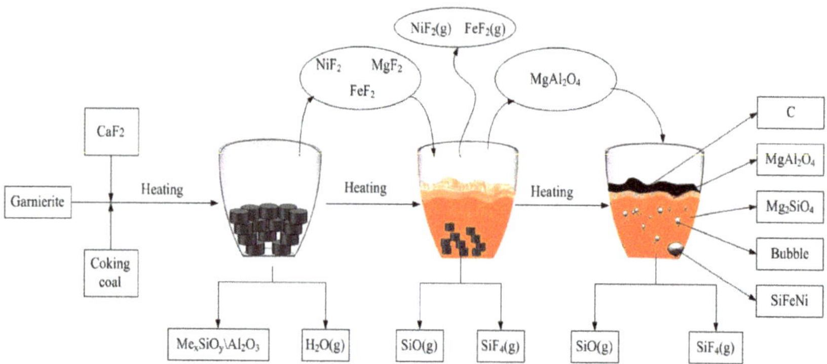

Figure 19. The influence and behavior of CaF_2 on the reduction process.

3.6. Magnetic Separation Process

Table 4 shows that after the magnetic separation, the grades of Fe were low, probably because Fe was present in the forsterite and spinel during the reduction process, which resulted in a low chemical analysis result. At the optimal conditions, the direct recoveries of Fe and Ni were 82.97% and 98.21%, respectively, in the vacuum carbothermal reduction–magnetic separation process, and the enrichment ratios were 3.18 and 9.35, respectively, as shown in Table 5. The data in the tables indicate that adding a proper amount of CaF_2 could effectively improve the grades and recovery of Fe and Ni.

Table 4. The content and recovery of Fe and Ni using the magnetic separation process and using the vacuum carbothermal reduction–magnetic separation process.

Temperature (K)	CaF_2 (wt. %)	Quality of Magnetic Material (g)	Chemical Analysis of Fe, Ni in Magnetic Materials (%)		Quality of Non-Magnetic Material (g)	Chemical Analysis of Fe, Ni in Non-Magnetic Materials (%)	
			Fe	Ni		Fe	Ni
1623	3	24.49	32.46	5.18	35.66	0.42	0.04
1823	3	30.29	38.17	5.99	21.71	0.33	0.05
1723	0	50.54	14.95	1.04	50.54	0.00	0.00
1723	3	11.16	40.20	6.73	21.75	0.18	0.03
1723	6	23.61	38.30	6.69	50.57	0.23	0.07
1723	9	23.06	37.30	6.22	49.73	0.26	0.06
1723	12	19.03	34.75	5.11	49.13	0.38	0.02

Table 5. Enrichment ratio of Fe and Ni with different dosages of CaF_2 at different temperature and Recovery of Fe and Ni during the vacuum carbothermal reduction–magnetic separation process.

Temperature (K)	1623	1823	1723	1723	1723	1723	1723
CaF_2 (wt %)	3	3	0	3	6	9	12
Enrichment ratio of Fe (%)	2.56	3.02	1.18	3.18	3.03	2.95	2.75
Enrichment ratio of Ni (%)	7.19	8.32	1.44	9.35	9.29	8.64	7.10
Recovery of Fe (%)	73.19	73.87	71.50	82.97	54.03	51.55	40.90
Recovery of Ni (%)	84.89	93.62	87.45	98.21	81.19	64.64	48.69

4. Conclusions

A vacuum carbothermal reduction–magnetic separation method to recover Ni, Fe, and Mg from garnierite was proposed, and the experiments indicated that the optimal conditions for this method for the recovery of Fe and Ni are 1723 K, a molar ratio of ore/C of 1:1.2, and the addition of CaF_2 at 3%. The recovery of Fe and Ni reached 82.97% and 98.21%, respectively, in the vacuum carbothermal reduction–magnetic separation process, and the enrichment ratios of Fe and Ni reached a maximum of 3.18 and 9.35, respectively.

The CaF_2 in the vacuum carbothermal process formed low-melting-point compounds (FeF_2 and NiF_2) at 1315 K and 1400 K, respectively, to transform the raw materials from solid to liquid, which accelerated the mass transfer and the heat transfer of Fe–Ni particles, and formed Si–Ni–Fe alloy particles with diameters of approximately of 20 mm. Simultaneously, the smelting materials appeared stratified, hindering the reduction of magnesium.

The removal rate of Mg increased and the maximal recovery only reached 74.12%. The reason for this low Mg removal rate is the stratification of the coking coal and the molten materials, which limited the increase in the recovery of Mg. However, the recovery of Si and the removal rate of Mg showed the opposite trend, this becomes a contradiction point for recovering garnierite comprehensively. Previously, we used this method to recover Ni, Fe, Si, and Mg with CaO as an additive, achieved some good experiment results, especially for the removal rate of Mg [29]. Subsequently, we could consider the mixture of CaF_2–CaO as an additive to solve the contradiction of Si and Mg, and achieve the purpose of comprehensive recovery of valuable metals from garnierite.

The method used in this article can solve the problem of the large amounts of slag produced and the inability to recycle the magnesium in the traditional pyrometallurgical process. For the future industrial application, it is necessary to solve the problem of magnesium condensation technology. If magnesium can be recycled in a reduction process, not only can it increase the economic benefits of enterprises, but it can also solve the problem of environment pollution, therefore, it has a certain industrialization significance. In addition, this article clarifies the behavior mechanism of CaF_2 in the reduction process of garnierite, it can be used for reference in guiding industrial production.

Author Contributions: Conceptualization, T.Q.; writing—original draft preparation, Q.W.; writing—review and editing, X.G., M.L. and L.S.; funding acquisition, B.Y. and Y.D. All authors have read and agree to the published version of the manuscript.

Funding: This work was funded by the National Nature Science Foundation of China (No. 51604133) and the Academician Free Exploration Fund of Yunnan Province, China (No. 2019HA006).

Conflicts of Interest: The authors declare no conflict of interest.

References

1. Esmaeili, N.; Ojo, O.A. Analysis of Brazing Effect on Hot Corrosion Behavior of a Nickel-Based Aerospace Superalloy. *Metall. Mater. Trans. B* **2018**, *49*, 912–918. [CrossRef]
2. Zhang, L.N.; Szpunar, J.A.; Dong, J.X.; Ojo, O.A.; Wang, X. Dependence of Crystallographic Orientation on Pitting Corrosion Behavior of Ni-Fe-Cr Alloy 028. *Metall. Mater. Trans. B* **2018**, *49*, 919–925. [CrossRef]
3. Saad, E.A.; Hassanien, M.M.; Ellban, F.W. Nickel(II) diacetyl monoxime-2-pyridyl hydrazone complex can inhibit Ehrlich solid tumor growth in mice: A potential new antitumor drug. *Biochem. Biophys. Res. Commun.* **2017**, *484*, 579–585. [CrossRef] [PubMed]
4. Nasab, M.B.; Hassan, M.R. Metallic Biomaterials of Knee and Hip—A Review. *Trends Biomater. Artif. Organs* **2010**, *24*, 69–82.
5. Babić, M.M.; Božić, B.Đ.; Božić, B.Đ.; Filipović, J.M.; Ušćumlić, G.S.; Tomić, S.L. Evaluation of novel antiproliferative controlled drug delivery system based on poly(2-hydroxypropyl acrylate/itaconic acid) hydrogels and nickel complex with Oxaprozin. *Mater. Lett.* **2016**, *163*, 214–217. [CrossRef]
6. Setcos, J.C.; Mahani, A.B.; Silvio, L.D.; Mjor, I.A.; Wilson, N.H.F. The safety of nickel containing dental alloys. *Dent. Mater.* **2006**, *22*, 1163–1168. [CrossRef] [PubMed]
7. Zhang, L.; Yang, H.P.; Cao, F.; Feng, A.S.; Zhao, J.W. Study on Current Situation and Analysis of Supply and Demand of Global Nickel Resource. *Conser. Utilizat. Miner. Resour.* **2016**, *1*, 64–69.
8. Zong, K.; Zhou, J.Y.; Fu, S.X. Study on global market allocation of Chinese nickel metal resources. *Miner. Explor.* **2015**, *6*, 86–91.
9. Valix, M.; Tang, J.Y.; Cheung, W.H. The effects of mineralogy on the biological leaching of nickel laterite ores. *Miner. Eng.* **2001**, *14*, 1629–1635. [CrossRef]
10. Myagkiy, A.; Truche, L.; Cathelineau, M.; Golfier, F. Revealing the conditions of Ni mineralization in the laterite profiles of New Caledonia: Insights from reactive geochemical transport modelling. *Chem. Geol.* **2017**, *466*, 274–284. [CrossRef]
11. Alfantazi, A.M.; Moskalyk, R.R. Nickel laterite processing and electrowinning practice. *Miner. Eng.* **2002**, *15*, 593–605.
12. Ma, B.Z.; Wang, C.Y.; Yang, W.J.; Chen, Y.Q.; Yang, B. Comprehensive utilization of Philippine laterite ore, part 1: Design of technical route and classification of the initial ore based on mineralogical analysis. *Int. J. Miner. Process.* **2013**, *124*, 42–49. [CrossRef]
13. Khoo, J.Z.; Haque, N.; Bhattacharya, S. Process simulation and exergy analysis of two nickel laterite processing technologies. *Int. J. Miner. Process.* **2017**, *161*, 83–93. [CrossRef]
14. Elliott, R.; Pickles, C.A.; Forster, J. Thermodynamics of the Reduction Roasting of Nickeliferous Laterite Ores. *J. Miner. Mater. Charact. Eng.* **2016**, *4*, 320–346. [CrossRef]
15. Kursunoglu, S.; Ichlas, Z.T.; Kaya, M. Solvent extraction process for the recovery of nickel and cobalt from Caldag laterite leach solution: The first bench scale study. *Hydrometallurgy* **2017**, *169*, 135–141. [CrossRef]

16. Leonardou, S.A.; Tsakiridis, P.E.; Oustadakis, P.; Karidakis, T.; Katsiapi, A. Hydrometallurgical process for the separation and recovery of nickel from sulphate heap leach liquor of nickeliferrous laterite ores. *Miner. Eng.* **2009**, *22*, 1181–1192. [CrossRef]

17. Ma, B.Z.; Wang, C.Y.; Yang, W.J.; Yang, B.; Zhang, Y.L. Selective pressure leaching of Fe (II)-rich limonitic laterite ores from Indonesia using nitric acid. *Miner. Eng.* **2013**, *45*, 151–158. [CrossRef]

18. Kaya, S.; Topkaya, Y.A. High pressure acid leaching of a refractory lateritic nickel ore. *Miner. Eng.* **2011**, *24*, 1188–1197. [CrossRef]

19. Johnson, J.A.; Cashmore, B.C.; Hockridge, R.J. Optimisation of nickel extraction from laterite ores by high pressure acid leaching with addition of sodium sulphate. *Miner. Eng.* **2005**, *18*, 1297–1303. [CrossRef]

20. Muir, B.I.; Whittington, D. Pressure acid leaching of nickel laterites: A review. *Miner. Process. Extr. Metall. Rev.* **2000**, *21*, 527–599.

21. Dodbiba, G.; Kim, J.; Tanno, H.; Okaya, K.; Matsuo, S.; Fujita, T. Calcination of low-grade laterite for concentration of Ni by magnetic separation. *Miner. Eng.* **2010**, *23*, 282–288.

22. Li, G.H.; Shi, T.M.; Rao, M.J.; Zhang, Y.B. Beneficiation of nickeliferous laterite by reduction roasting in the presence of sodium sulfate. *Miner. Eng.* **2012**, *32*, 19–26. [CrossRef]

23. Zhu, D.Q.; Cui, Y.; Vining, K.; Hapugoda, S.; Douglas, J.; Pan, J.; Zheng, G.L. Upgrading low nickel content laterite ores using selective reduction followed by magnetic separation. *Int. J. Miner. Process.* **2012**, *106–109*, 1–7. [CrossRef]

24. Ma, B.Z.; Xing, P.; Yang, W.J.; Wang, C.Y.; Chen, Y.Q.; Wang, H. Solid-State Metalized Reduction of Magnesium-Rich Low-Nickel Oxide Ores Using Coal as the Reductant Based on Thermodynamic Analysis. *Metall. Mater. Trans. B* **2017**, *48*, 2037–2046. [CrossRef]

25. Cao, C.; Xue, Z.L.; Duan, H.J. Making Ferronickel from Laterite Nickel Ore by Coal-Based Self-Reduction and High Temperature Melting Process. *Int. J. Nonfer. Metall.* **2016**, *5*, 9–15. [CrossRef]

26. Li, J.H.; Chen, Z.F.; Shen, B.P.; Xu, Z.F.; Zhang, Y.F. The extraction of valuable metals and phase transformation and formation mechanism in roasting-water leaching process of laterite with ammonium sulfate. *J. Clean. Prod.* **2017**, *140*, 1148–1155. [CrossRef]

27. Xu, C.; Cheng, H.W.; Li, G.S.; Lu, C.Y.; Lu, X.G.; Zou, X.L.; Xu, Q. Extraction of metals from complex sulfide nickel concentrates by low-temperature chlorination roasting and water leaching. *Int. J. Miner. Metall.* **2017**, *24*, 377–385. [CrossRef]

28. Zhou, S.W.; Wei, Y.G.; Li, B.; Ma, B.Z.; Wang, C.Y.; Wang, H. Kinetics study on the dehydroxylation and phase transformation of $Mg_3Si_2O_5(OH)_4$. *J. Alloys Compd.* **2017**, *713*, 180–186. [CrossRef]

29. Wang, Q.; Qu, T.; Gu, X.P.; Shi, L.; Yang, B.; Dai, Y.N. The effect of CaO on the recovery of Fe and Ni in a vacuum carbothermal reduction of garnierite. *J. Min. Metall. Sect. B Metall.* **2019**, *55*, 351–358. [CrossRef]

Article

Metallothermic Al-Sc Co-Reduction by Vacuum Induction Melting Using Ca

Frederic Brinkmann, Carolin Mazurek and Bernd Friedrich

IME Process Metallurgy and Metal Recycling, RWTH Aachen University, 52056 Aachen, Germany;
cmazurek@metallurgie.rwth-aachen.de (C.M.); bfriedrich@metallurgie.rwth-aachen.de (B.F.)

* Correspondence: fbrinkmann@metallurgie.rwth-aachen.de; Tel.: +49-241-8095196
† Current address: IME Process Metallurgy and Metal Recycling, Intzestr. 3, 52072 Aachen, Germany.

Received: 9 October 2019; Accepted: 10 November 2019; Published: 14 November 2019

Abstract: Due to its enhancing properties in high-tech material applications, the rare earth element Scandium (Sc) is continuously gaining interest from researchers and material developers. The aim of this research is to establish an energy and resource efficient process scheme for an in situ extraction of Al-Sc master alloys, which offers usable products for the metallurgical industry. An AlSc20 alloy is targeted with an oxyfluoridic slag as a usable by-product. The thermochemical baseline is presented by modelling using the software tool FactSage; the experimental metal extraction is conducted in a vacuum induction furnace with various parameters, whereas kinetic aspects are investigated by thermogravimetric analysis. The Sc-containing products are analyzed by ICP-OES/IC concerning their chemical composition. Optimum parameters are derived from a statistical evaluation of the Sc content in the obtained slag phase. The material obtained was high in Ta due to the crucible material and remarkably low in Al and F; a comparison between the modelled and the obtained phases indicates kinetic effects inhibiting the accomplishment of equilibrium conditions. The formation of a Sc-rich Al-Sc phase (32.5 wt.-% Sc) is detected by SEM-EDS analysis of the metal phase. An in situ extraction of Al from Ca with subsequent metallothermic reduction of ScF_3 as a process controlling mechanism is presumed.

Keywords: scandium; master alloys; aluminum alloys; metallothermy; vacuum induction melting; factsage

---

1. Introduction

Material scientists have highlighted the remarkable properties of Sc as an alloying element for Al(-Mg) for a long time [1–4]. It is proven to have a positive impact on the strength even at small concentrations below 0.5 wt.-%, as well as a raising effect on the recrystallization temperature, which directly leads to enhanced weldability and hot cracking behavior of the material. For Al alloys, the properties are linked to an intermetallic compound Al_3Sc that shows little misfit with the Al matrix. However, extraction of Sc from its fluoridic or oxidic form to form a metal phase is challenging due to its high affinity to oxygen and halogenides, its high melting point, and also its low density. Special processes that have been used for the extraction are usually inefficient, costly, unscalable or are associated with low yields. Electrochemical processes may be applied, most notably as molten salt electrolysis. The principles of Hall–Heroult and the FCC process are under investigation for the application in Sc metallurgy. Low-temperature electrolysis in ionic liquids are of high potential but are not applicable for large production quantities, rather for applications such as coatings. Processing via carbothermic reduction does not provide a fully reduced product [5]. By employing the techniques of vacuum induction melting (VIM), the parameters are easy to manipulate, reaction times are fast, and the input material may be versatile. The application of reduction processes via VIM is unusual but has been performed before [6].

In 1937, the extraction of Sc metal from Sc precursors was first reported by Fischer et al. [7]. Motivated by the interest in the physical and chemical properties of the rare earth element, calciothermy was applied to reduce ScF_3. At this point in time, tantalum crucibles or sheets were not available; instead, oxidic crucibles were used, which led to limited success in the product purity. The principles were later advanced by Spedding et al. [8], who used Ta crucibles for the reduction processes and subsequent refining steps, which resulted in 10 g Sc with minor impurities.

At the same time, research on the in situ production of Al-Sc master alloys by aluminothermic reduction of Sc salts has been studied by numerous research with increasing numbers of publications within the last few years. Most research has been carried out on master alloy production in the range of 2–10% Sc. The approach for most experimental campaigns is to dissolve ScF_3 in a suitable salt slag and to induce the reduction by bringing it in contact with Al [9]. The strong advantage of this technology is the high Sc conversion yield (up to 91% for Al-Sc 2%) as well as the prevention of harmful off-gases. On the other hand, only low Sc concentrations may be incorporated in the Al melt and the resulting slag residue phase has little economic value. By using pure ScF_3 and Al as input material, higher Sc contents in the Al matrix are achievable. Data in the literature indicate that the maximum Sc content is governed by the process temperature—at a maximum temperature of 1040 °C, Sc content reached 10% [10]. However, processing with this system leads to the formation of fluoridic gases, mainly AlF and AlF_3. A thorough analysis of the sublimates is given by Sokolova et al. [11]. Used refractory materials are usually graphite [9,10] or Ta [8,12]. MgO refractory material was used at the beginning of the research on Sc reduction [7]. Furnaces used for the reduction step are induction furnaces in open atmosphere [9], a retort type reactor [10], and resistance heating furnaces [11,12]. Hydrofluoric acid is used for ScF_3 synthesis in the classical approach. Since the handling of HF is implicating safety issues and an additional processing step is involved, the Sc extraction from Sc_2O_3 is desired. Thus, metallothermic reduction of the oxide is thermochemically unfeasible for pure substances on a technological basis. Therefore, different approaches are described in the literature that mostly comprise in situ conversion of the oxide to fluoride. Harata et al. [12] used a mixture of Al and Ca as reducing agent and produced an alloy with roughly 9 wt.-% Sc in the Al matrix, indicating a complete conversion of the oxide into the metallic form.

A novel approach for in situ extraction of Al-Sc master alloys is presented in the present paper. By combining the reduction of Al and Sc in one step, the alloying of Sc to Al melts, which are often paired with Sc losses in a dross phase, is surpassed. The by-product, a slag phase with a liquidus temperature of <1400 °C, enables an all-liquid system that is treatable in conventional vacuum induction furnaces. The use of a Ta crucible allows fast heating rates and stability against the formed liquid phases. Reduction mechanism and pathways are proposed as a result of a thermogravimetric analysis and factorial design of experiment based VIM experiments.

2. Materials and Methods

2.1. Thermochemical Considerations

The thermochemical software tool FactSage^{TM} (7.2, Thermfact/CRCT, Montreal, Canada/ GTT-Technologies, Aachen, Germany) was used to examine phase equilibria in the metallothermic systems. The main ambition is to track changes within the systems with varying temperatures and compositions regarding the addition of reducing agents. For the calculations of phase diagrams, the module "Phase Diagram" was used, while equilibrium calculations were performed in "Equilib". The databases considered were FactPS, FTlite, FToxide, and FTsalt. Supplementary Materials is provided containing information on the settings used for the calculations.

A novel approach to produce an in situ Al-Sc master alloy is to reduce both precursors of Al (Al_2O_3) and Sc (ScF_3) by Ca simultaneously, using the advantage of additional reaction enthalpy by Al extraction that boosts the reduction process and giving the opportunity to promote the formation of a lower temperature liquid CaF_2-Al_2O_3-CaO-(ScF_3) slag as expressed for a complete reduction of Al_2O_3 and ScF_3 in Equation (1):

$$(\frac{3}{2}x + 3y)Ca + xScF_3 + yAl_2O_3 = Al_{\frac{y}{2}}Sc_x + (\frac{3}{2}x + 3y)(CaO \cdot CaF_2). \tag{1}$$

In the system observed, fluoridic compounds are present that often exhibit high vapor pressures. Here, ScF_3, CaF_2, AlF_3, and AlF are most likely to evolve as gaseous species due to their relatively high vapor pressures. Hence, losses of Al and F are expected in a pure substances scenario. Furthermore, a reaction between Al and CaF_2 is observed via simulation, indicating a conversion at elevated temperatures leading to enhanced AlF formation (Figure 1).

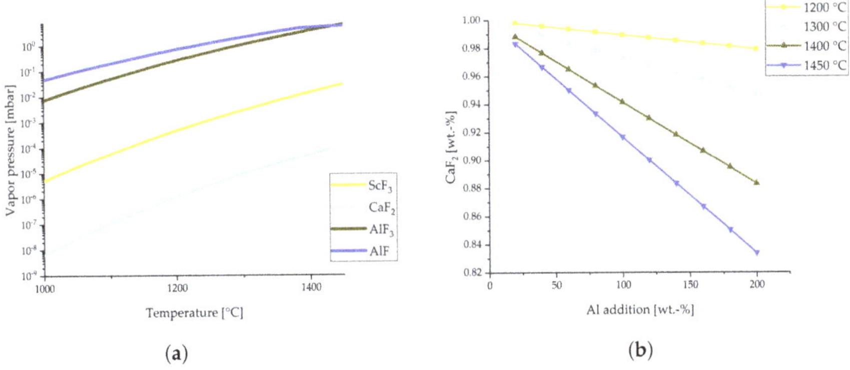

Figure 1. (**a**) vapor pressures of selected fluoridic compounds; (**b**) calculated CaF_2 stability relative to the weight fraction of Al added at different temperatures.

A thorough investigation of the most stable phases formed during variation of process parameters for Equation (1) is shown here with special emphasis on metal and slag phases.

2.1.1. Al-Sc Alloys

The resulting metal phase from the co-reduction of Equation (1) is composed of Al, Sc, and C with neglectable F and O contents. For a binary Al-Sc approach, with increasing Sc content in Al, the liquidus temperature is increased with the formation of Al-Sc intermetallic phases. At Sc contents of 35.7 wt.-%, the compound Al_3Sc is formed peritectically, whereas Al_2Sc (Laves phase) and AlSc are occurring at higher Sc concentrations.

An efficient extraction process ensures a full conversion of Sc into its metallic form. However, it is also the most ignoble element in the system, considering only pure substances. In a ternary solution, Sc activity is strongly depending on the composition and may be lowered below a certain threshold where it is inactive as reducing agent. The calculated Sc activity in the Al-rich corner of the Al-Sc-Ca ternary phase diagram is depicted in Figure 2. Note that Ca has a strong effect on the activity of Sc at concentrations exceeding 10 wt.-%. The critical activity is determined as the threshold below which Al_2O_3 as a pure substance is more stable than Sc_2O_3 and is found to be at 3.65×10^{-5}. In the absence of Ca, 16 wt.-% Sc in Al are found to be the highest concentration feasible. Hence, controlling the Ca concentration is highly important for optimal Sc yields. At 1400 °C, the elemental concentrations of Sc, Al and Ca as computed are listed in Table 1.

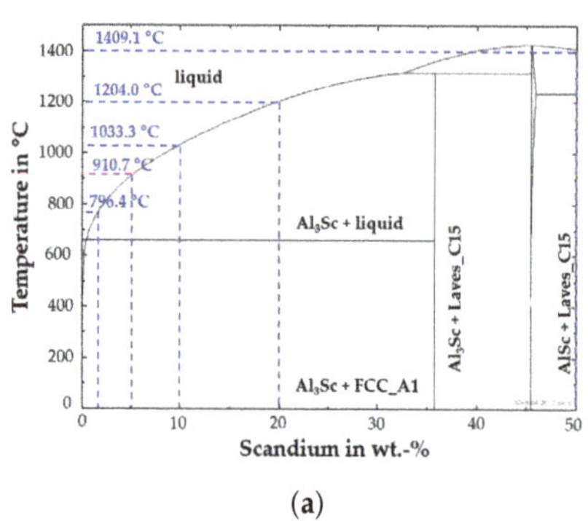
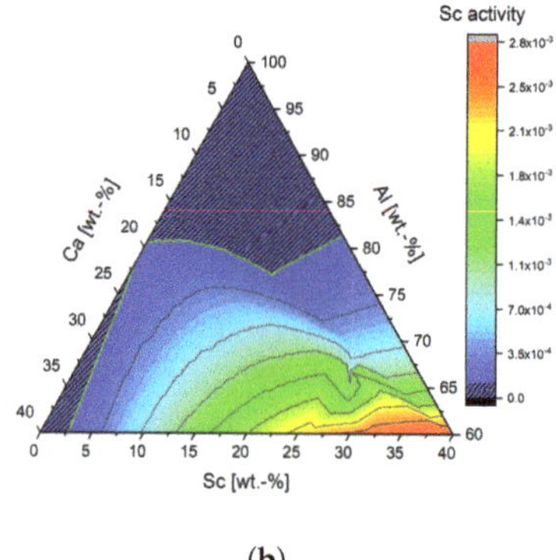

Figure 2. (**a**) calculated Al-Sc binary phase diagram up to 50 wt.-% Sc; (**b**) calculated Sc activity in the Al-rich corner of the ternary Al-Sc-Ca system at 1400 °C. The highlighted region indicates the range in which Sc activity is below the critical threshold.

Table 1. Sc, Al, and Ca concentrations calculated at different target ratios in metal phase at 1400 °C.

Targeted Sc Concentrations in wt.-%	Sc in wt.-%	Al in wt.-%	Ca in wt.-%
2	1.8	75.9	22.3
5	4.4	74.3	21.3
10	9.0	72.2	18.9
20	18.5	68.2	13.3
50	42.8	49.0	8.3

2.1.2. Slag Design

The ternary phase diagram of CaO-CaF_2-Al_2O_3 was examined to predict slag behavior originating from the reduction. The boundary systems CaO-Al_2O_3 and CaF_2-CaO are stable, whereas CaF_2-Al_2O_3 may decompose. Above its sublimation temperature, the evolving AlF_3 volatizes, thus disabling equilibria conditions in an open system. In low CaO activity, CaF_2-Al_2O_3 slags, AlF_3 evaporation is observed by Shinmei and Machida [13]. The underlying mechanism for the gas formation being a reaction of AlF_3 with Al was proven by Dyke et al. [14].

The oxyfluoridic by-product of the Al-Sc co-reduction exhibits a high melting point above 2000 °C, if a stoichiometric conversion is processed. Hence, it is mandatory to add fluxing compounds that lower the liquidus temperature and do not contaminate the metal phase. By adding CaF_2 and Al_2O_3 at a ratio of 7:3, slag liquidus temperatures as low as the projected melting point of the alloys (1200 °C) may be realized (Figure 3). Note that the calculation does not include the Sc-bearing compounds that might be dissolved in the slag, due to the assumption of a complete Sc reduction. In addition, metal dissolution is not considered here as an effective parameter on the liquidus temperature.

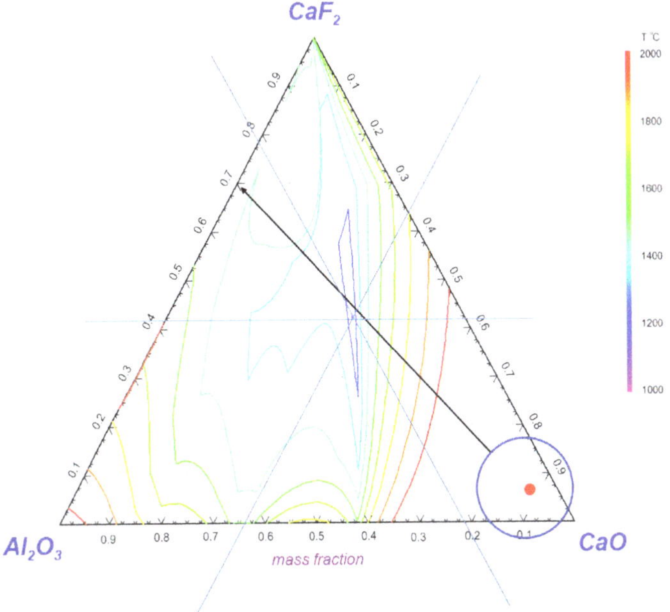

Figure 3. Ternary oxyfluoridic phase diagram CaF$_2$-Al$_2$O$_3$-CaO with isotherms. The arrow depicts the slag design strategy to minimize the liquidus temperature originating from the non-fluxed system as indicated by the dot.

2.2. Materials

As basic material mixture, the thermochemical considerations were used to determine the ratios at which

- Sc extraction yield is at 100%,
- The resulting slag phase has a liquidus temperature of <1400 °C,
- Minimum Ca contents are expected in the metallic matrix.

ScF$_3$ as the Sc bearing compound is fully reduced when stiochiometric amounts of Ca are present. Furthermore, the surplus Ca is converted by reducing Al$_2$O$_3$. Hence, CaO is formed which is combining with Al$_2$O$_3$ and CaF$_2$ to a liquid slag. Ultimately, the following ratios are employed (Table 2):

Table 2. Input ratio of materials.

Compound	ScF$_3$	Al$_2$O$_3$	CaF$_2$	Ca
Weight factor	1.00	6.33	7.00	4.52

At this ratio, liquidus temperature of the slag produced is 1247.5 °C. The slag has a moderate viscosity in comparison to slag without additive addition (viscosity calculation with FactSage). In addition, an Al-Sc alloy with 20 wt.-% Sc is the result of the thermochemical simulation with a liquidus temperature of the alloy of 1204.0 °C and a modest Ca contamination.

ScF$_3$ was analyzed via XRF in order to identify impurities that might be critical for the subsequent reduction process. Thus, material analysis via ICP-OES/IC was discarded in that matter as an overview of the impurities was the main purpose. ScF$_3$ purity was determined to be 99.584% with minor Na, Al, and Si impurities (Table 3). Ca shots (SigmaAldrich, 99 wt.-% granular) were used as reducing agent, Al$_2$O$_3$ (99.98 wt.-% powder, milled for 2 min), and CaF$_2$ (Alfa Aesar, 99 wt.-% powder) as fluxes.

Table 3. XRF analysis of ScF_3.

ScF_3	Na_2O	Al_2O_3	SiO_2	P_2O_5	CaO	Fe_2O_3	CuO	SO_3
99.584%	0.113%	0.122%	0.047%	0.005%	0.014%	0.021%	0.013%	0.080%

A distribution of particle sizes was determined by sieving analysis (Retsch Vibratory Sieve Shaker AS 200 control, Haan, Germany). For the ScF_3 used, fine as well as coarse particles were found, indicating a broad particle size spectrum. While CaF_2 is present at fine particle sizes with 75% smaller 45 μm, Al_2O_3 was found to be rather large, with a greater share above 63 μm. Ca shots were also sieved, with half of the weight above 2.5 mm and only a small fraction of granules below 2 mm (14%). The results of the screen analysis performed on the input material are shown in Table 4.

Table 4. Particle size distribution of input materials.

Material	>90 μm	<90 μm	<63 μm	<45 μm	*total*
ScF_3	37%	7%	14%	37%	95%
CaF_2	11%	6%	2%	75%	94%
Al_2O_3	48%	37%	7%	1%	93%
	>2.5 mm	<2.5 mm	<2 mm		
Ca	54%	31%	14%		

2.3. Experimental Procedure

The thermogravimetric analysis (TG) was performed on a thermobalance (Netzsch STA 409, alumina sample cups, 3.4 mL volume, Selb, Germany) at a heating rate of $10 \, K \cdot min^{-1}$ and a maximum temperature of 1500 °C. Argon was flushed continuously at $150 \, mL \cdot min^{-1}$. A triple Yttria-coating was applied onto the crucibles' interior in order to avoid any reaction with the material.

A laboratory scale vacuum induction furnace from Leybold–Heraeus with a nominal power of up to 5 kW, 10 kHz frequency, and a max. melting capacity of 60 mL is used for the extraction experiments. Inside the water-cooled induction coil, a Ta crucible is placed, which is coupling with the induction field to enable heat control of the system. A ZrO_2 isolation layer is stabilizing the crucible, together with refractory wool on the surface of the coil. The material is mixed prior to melting and is placed inside the crucible; the system is covered by a Nb condenser unit to minimize dust distribution within the vacuum chamber and to collect condensed material. Temperatures are measured continuously via three type B (Pt-PtRh) thermocouples inside the bulk (Nb protection tube), as well as on the condensator bottom and top (Figure 4). Fifteen experiments were performed with process temperature, Ar pressure, retention time, and Ca surplus variation on the basis of 4 g ScF_3. The alternated values are listed in Table 5.

Table 5. Parameters varied for the factorial design of experiment.

Temperature in °C	Pressure in mbar	Retention Time in min	Ca Surplus in %
1200–1450	200–800	15–45	100–200

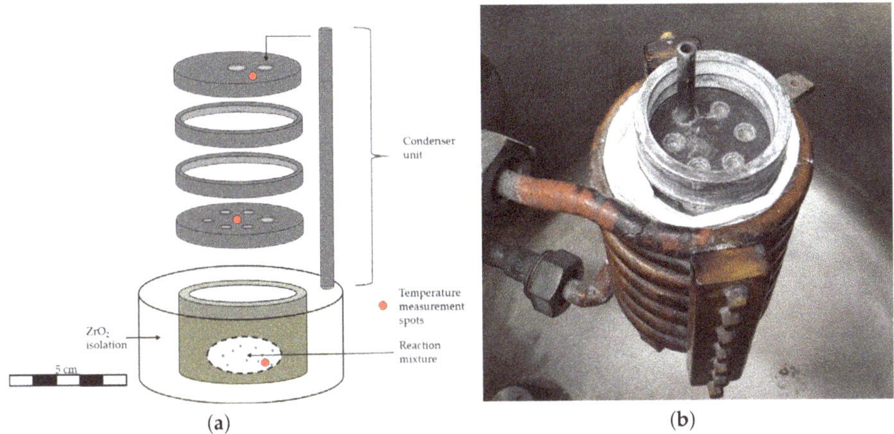

(**a**) (**b**)

Figure 4. (**a**) schematic setup inside the induction coil; (**b**) real setup viewed from above without a condenser top.

3. Results

3.1. Thermogravimetric Analysis

TG analysis of the co-reduction mixtures indicate an initial weight loss occurring at 960 °C, independent of CaF_2 fluxing. For the stochiometric co-reduction without CaF_2, the weight loss rate is relatively constant over the temperature range of 960 °C to 1360 °C. Above that, the weight loss rate decreases slightly. A different behavior is observed when CaF_2 is added; here, the mass loss decelerates at 1225 °C before increasing at 1410 °C (Figure 5).

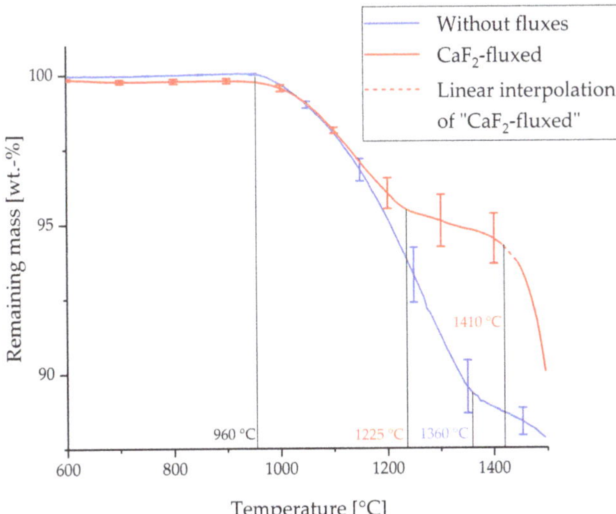

Figure 5. TG analysis of the co-reduction mixture without fluxing (blue) and with CaF_2 as flux (red).

3.2. Al-Sc Synthesis via VIM

The Sc content in the slag was found to be higher than expected, with 2.12 wt.-% on average. With Al and Ca corresponding well to the calculated versus the experimental concentrations, F is present at far lower contents than expected by thermochemical modelling. A slight Ta pick-up from the crucible is also observed. On the other hand, Al and Sc concentrations in the metal machined from the crucible are far below the expected range, whereas Ca and Ta contribute mostly to the metallic phase (Figure 6).

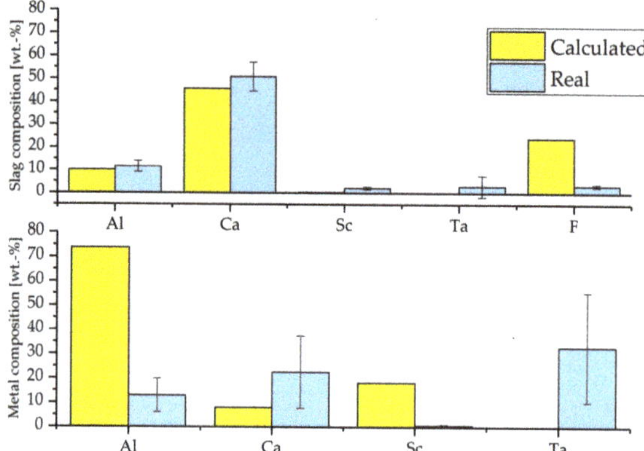

Figure 6. Mean ICP-OES analysis values of the slag and metal phases obtained; IC analysis for F.

The main effects contributing to the final Sc concentration of the oxyfluoridic phase are shown in Figure 7. Temperature dependency is indicating a minimum Sc slag content at 1330 °C; lower pressures are favorable for low Sc concentrations. While the retention time does not have a severe impact on the Sc content, high calcium surpluses have similar impact as low pressures.

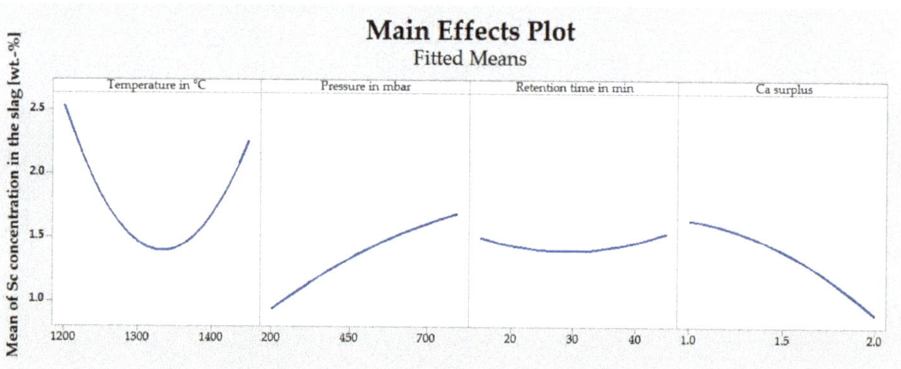

Figure 7. Main effects plot for the Sc content in the produced slag phase.

Slag and metal samples of the experiment with the parameters closest to the found optimum (1450 °C, 200 mbar, 15 min retention time, 200% Ca) were analyzed by SEM-EDS and are depicted in Figures 8 and 9. Elemental mapping of the metal phase shows a Sc-rich region clearly separated from Ca, whereas Al is found conjoined with Sc and Ca. Ta fragments originating from the crucible are visible as well. The measured spectrum in the Sc-rich region is consisting mostly of Al (59.2 wt.-%) and Sc (32.5 wt.-%), respectively. Thus, this region corresponds well to the composition Al_3Sc if minor impurities of Ca, Fa and Ta are discarded.

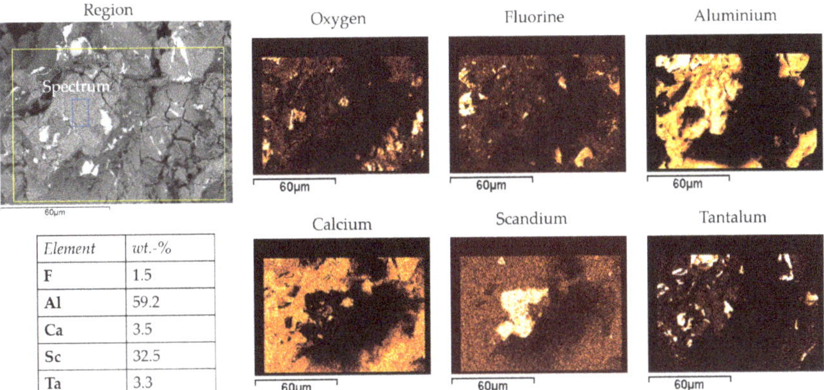

Figure 8. Metal phase SEM-EDS mapping and spectral analysis on O, F, Al, Ca, Sc and Ta.

For the slag samples, the mapping indicates the formation of an oxyfluoridic slag as O and F are found with similar intensities in the same regions. The slag is containing Al and Ca, as well as minor Sc and Ta concentrations, where the latter may be attributed to crucible wearing. Regions with higher Ca and F concentrations are found that are corresponding to the presence of CaF_2.

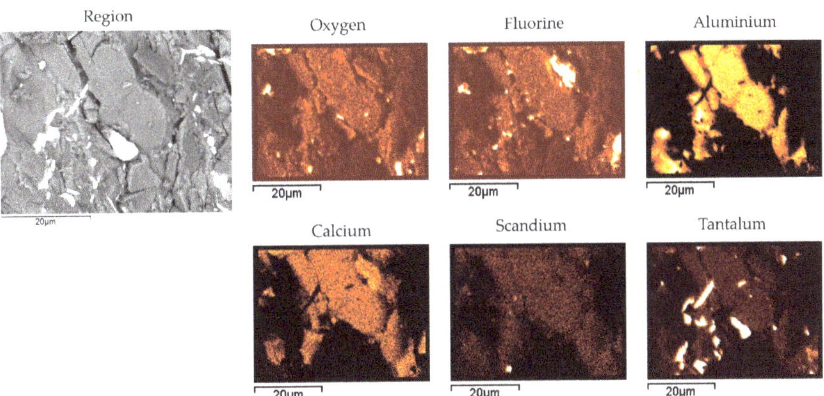

Figure 9. Slag phase SEM-EDS mapping on O, F, Al, Ca, Sc and Ta.

3.3. Interdependencies of Parameters

Over the temperature range investigated, a decrease in pressure is equally favorable at all temperatures. On the other hand, the retention time does not seem to have an impact on the extraction efficiency at different temperatures, whereas an increase in Ca addition is especially effective at low temperatures at 1200 °C. No interdependency is observed between pressure and retention time. Nevertheless, the application of low pressures is substantially supported by increasing Ca additions.

4. Discussion

The thermogravimetric investigations that were conducted in order to obtain insights into the reaction mechanisms during the calciothermic co-reduction of Al_2O_3 and ScF_3 indicated temperature ranges in which the processes occur. Especially between 960 °C and 1390 °C, mass losses are observed. It is assumed that initial weight losses may be attributed to the evaporation of Ca that trigger the reduction of the present Sc and Al precursors. The weight loss is decreasing above 1390 °C, indicating a completion of the reaction process. However, if fluxing agents (additional Al_2O_3 and CaF_2) are added, a temperature range between 1225 °C and 1410 °C is observed in which weight loss is almost stopped. Here, kinetic effects might hinder the evolving Ca; at higher temperatures, the weight loss increases dramatically, which may be attributed to the evaporation of fluoridic compounds, such as CaF_2.

Vacuum induction melting of the metallothermic mixtures was conducted with regard to the parameters' temperature, retention time, Ca addition, and pressure. The resulting slag may be extracted and analyzed for its Sc content that was thus chosen as target value in order to compare the influence of the parameters examined. Hence, the major findings of the study were as follows:

- The co-reduction of Al_2O_3 and ScF_3 with Ca results in the phase separation of a slag and a metal. Tantalum crucibles are not practicable due to the strong attack by the resulting metal phase.
- SEM-EDS analysis of the slag and metal samples confirms the synthesis of an oxyfluoridic slag; Al-Sc with 32.5 wt.-% Sc is found in fragments.
- In the chosen experimental setting, the contamination of the metal sample with Ta and Ca is too excessive to be used directly for alloying processes. However, the Sc concentration without impurities amounts to 4.68 wt.-% on average, which is above the concentration for commercial Al-Sc master alloys (2 wt.-%). A refining step needs to be investigated in order to remove the impurities by distillation/sublimation.
- Temperature, pressure, and retention time strongly affect the Sc content remaining in the slag.
- Gas–solid reactions between the evolving Ca and the feedstock are found to be the major reaction mechanism. This assumption is supported by the results from TG analysis, as well as the preference of low pressures (200 mbar) within the system.
- The low concentration of Al in the metal phase, as well as the F concentration that was found to be much smaller compared to the thermochemical calculations, which might indicate an evaporation of AlF_3 that was not taken into account within the thermochemical considerations.

The interdependencies of the parameters investigated enable the drawing of conclusions regarding the mechanisms of Al-Sc co-reduction. Due to the observation that the retention time does not have an impact on the Sc content at various temperatures applied, it is probable that the reduction is already complete at the minimum temperature of 1200 °C. With the decreasing impact of Ca addition at elevated temperatures, it becomes apparent that a mechanism counteracts the calciothermic reduction that occurs at lower temperatures. Thus, there is a retransfer of Sc into the slag phase that might be associated with the increasing evaporation observed also by TG.

While the pressure influence is relatively independent from the retention time, low pressures are very effective when Ca is present at twice the stochiometric need of the co-reduction. We assume that the governing Ar pressure is acting as a kinetic barrier that hinders the free flow of Ca gas through the bulk particles. Thus, Ca does not get in contact with the material and leaves it unreacted. The hypothesis is supported by the observation that higher pressures are not influenced by high Ca additions to the system, indicating a high ratio of unreacted Ca. Furthermore, the positive effect of increasing Ca addition on longer retention times indicate the suppression of a Sc retransfer mechanism when surplus Ca is present.

5. Conclusions

With regard to the differences between ICP/IC analysis and the calculated concentrations of Al and F especially, a mechanism for the removal of these elements from the system is the most probable explanation (Figure 10). Therefore, CaF_2 has to be converted into a different compound with F evaporation in a volatile compound. As shown in Figure 10, we propose a reaction path where volatized Ca gas reduces primarily Al_2O_3 at temperatures exceeding the evaporation point of Ca (Reaction step 1). Liquid Al is therefore available for subsequent reduction of ScF_3 (Reaction step 3) and the conversion of CaF_2 to CaO in conjunction with Al_2O_3 (Reaction step 6). At temperatures above the liquidus temperature of the slag (Reaction step 7), the formation of the liquid phases decelerates the evaporation rate substantially.

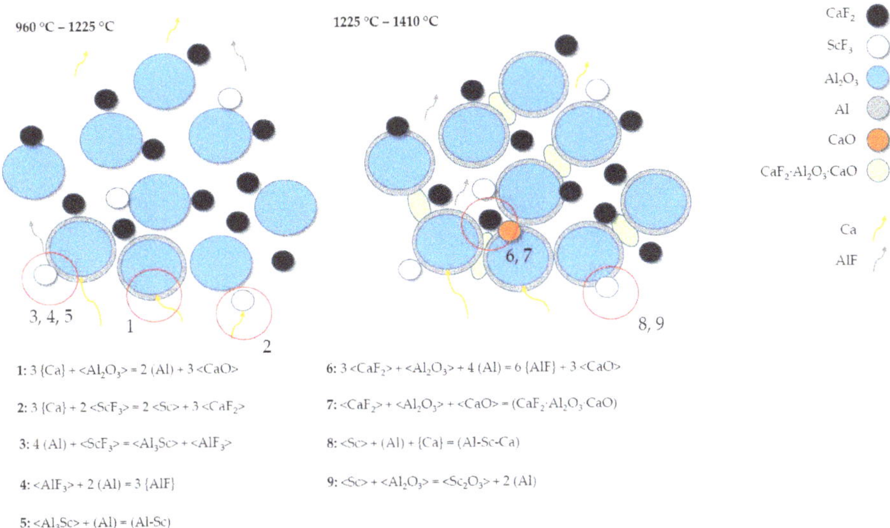

Figure 10. Proposed reaction mechanisms occurring during the co-reduction of ScF$_3$ and Al$_2$O$_3$ with Ca.

Further research on the topic needs to be performed regarding the thermochemical and kinetic evaluation of the system in the non-equilibrium state. Evaporation of fluoridic compounds is the main driver for deviations from calculated system behaviors that need to be taken into account for optimized process settings. Ultimately, a suitable container for the synthesized Al-Sc alloy needs to be chosen for a proper phase separation as Ta crucibles are highly reactive with the metal phase. Nevertheless, a refining step for Ca removal is necessary in order to reveal the potential of this technology with which Sc master alloy concentrations in suitable quantities for alloying processes are feasible.

Supplementary Materials: The following are available online at http://www.mdpi.com/2075-4701/9/11/1223/s1, Description: Modeling Metallothermic Reduction of Aluminium-Scandium Alloys in FactSage 7.2.

Author Contributions: Conzeptualization, F.B. and B.F.; methodology, F.B. and C.M.; investigation, F.B. and C.M.; writing–original draft preparation, F.B.; writing–review and editing, C.M. and B.F.; project administration, B.F.; funding acquisition, B.F.

Funding: This research was funded by the European Union's Horizon 2020 research and innovation program under the Grant Agreement No. 730105-SCALE.

Acknowledgments: The authors would like to thank Ulrike Hecht of Access e.V., Aachen, for the SEM-EDS analyses performed.

Conflicts of Interest: The authors declare no conflict of interest.

Abbreviations

The following abbreviations are used in this manuscript:

ICP-OES	Inductively Coupled Plasma Optical Emission Spectometry
IC	Ion Chromatography
SEM-EDS	Scanning Electron Microscopy-Energy Dispersive X-Ray Spectroscopy
FCC	Fray–Farthing–Chen–Cambridge
VIM	Vacuum Induction Melting
TG	Thermogravimetric

References

1. Lathabai, S.; Lloyd, P.G. The effect of scandium on the microstructure, mechanical properties and weldability of a cast Al-Mg alloy. *Acta Mater.* **2002**, *50*, 4275–4292. [CrossRef]
2. Parker, B.A.; Zhou, Z.F.; Nolle, P. The effect of small additions of scandium on the properties of aluminium alloys. *J. Mater. Sci.* **1995**, *30*, 452–458. [CrossRef]
3. Røyset, J.; Ryum, N. Scandium in aluminium alloys. *Int. Mater. Rev.* **2005**, *50*, 19–44. [CrossRef]
4. Zakharov, V.V.; Fisenko, I.A. Alloying Aluminum Alloys with Scandium. *Metal Sci. Heat Treat.* **2017**, *59*. [CrossRef]
5. Hfijek, B.; Karen, P.; Bro, V. Studies on Hydrolyzable Carbides. XXII: The Carbothermal Reduction of Scandium Oxide Sc_2O_3. *Monatshefte für Chemie* **1986**, *1278*, 1271–1278.
6. Halmann, M.; Steinfeld, A.; Epstein, M.; Vishnevetsky, I. Vacuum carbothermic reduction of alumina. *Miner. Process. Extr. Metall. Rev.* **2014**, *35*, 126–135. [CrossRef]
7. Fischer, W.; Brünger, K.; Grieneisen, H. Über das metallische Scandium. *Zeitschrift fur Anorganische und Allgemeine Chemie* **1937**, *231*, 54–62. [CrossRef]
8. Spedding, F.H.; Daane, A.H.; Wakefield, G.; Dennison, D.H. Preparation and Properties of High Purity Scandium Metal. *Trans. Metall. Soc. AIME* **1960**, *218*, 608–611.
9. Xu, C.; Liu, X.; Ma, F.; Wang, Z.; Wang, W.; Ma, C. Preparation Of Al-Sc Master Alloy By Aluminothermic Reaction With Special Molten Salt. In Proceedings of the 13th International Conference on Aluminum Alloys (ICAA13), TMS (The Minerals, Metals & Materials Society), Pittsburgh, PA, USA, 3–7 June 2012; pp. 195–200.
10. Mukhachov, A.P.; Kharitonova, E.A.; Skipochka, D.G. Scandium and its alloys with aluminum. *Probl. At. Sci. Technol.* **2016**, *101*, 45–50.
11. Sokolova, Y.V.; Pirozhenko, K.Y.; Makhov, S.V. Concentration of scandium during processing the sublimate of production of the aluminum-scandium master alloy. *Russ. J. Non-Ferrous Met.* **2015**, *56*, 10–14. [CrossRef]
12. Harata, M.; Nakamura, T.; Yakushiji, H.; Okabe, T.H. Production of scandium and Al-Sc alloy by metallothermic reduction. *Trans. Inst. Min. Metall. Sect. C Miner. Process. Extr. Metall.* **2008**, *117*, 95–99. [CrossRef]
13. Shinmei, M.; Machida, T. Vaporization of AlF_3 from the Slag CaF_2-Al_2O_3. *Metall. Trans.* **1973**, *4*, 1996–1997. [CrossRef]
14. Dyke, J.; Kirby, C.; Morris, A.; Gravenor, B. A Study of Aluminium Monofluoride and Aluminium Trifluoride by High-Temperature Photoelectron Spectroscopy. *Chem. Phys.* **1984**, *88*, 289–298. [CrossRef]

Article

Spectral Characterization of Copper and Iron Sulfide Combustion: A Multivariate Data Analysis Approach for Mineral Identification on the Blend

Walter Díaz [1], Carlos Toro [1,2,*], Eduardo Balladares [1], Victor Parra [1], Pablo Coelho [3], Gonzalo Reyes [1] and Roberto Parra [1]

[1] Metallurgical Engineering Department, Universidad de Concepción, Concepción CCP 4070386, Chile; walterdiaz@udec.cl (W.D.); eballada@udec.cl (E.B.); vparras@udec.cl (V.P.); gonzaloreyes@udec.cl (G.R.); rparra@udec.cl (R.P.)

[2] Dirección de Investigación, Universidad Tecnológica de Chile INACAP, Huechuraba, Avenida el Condor 720, Ciudad Empresarial, Huechuraba 8580000, Chile

[3] Facultad de Ingeniería y Tecnología, Universidad San Sebastián, Lientur 1457, Concepción 4080871, Chile; pablo.coelho@uss.cl

*   Correspondence: ctoron@inacap.cl; Tel.: (+56)-41-2928585

Received: 29 August 2019; Accepted: 16 September 2019; Published: 19 September 2019

Abstract: The pyrometallurgical processes for primary copper production have only off-line and time-demanding analytical techniques to characterize the in and out streams of the smelting and converting steps. Since these processes are highly exothermic, relevant process information could potentially be obtained from the visible and near-infrared radiation emitted to the environment. In this work, we apply spectral sensing and multivariate data analysis methodologies to identify and classify copper and iron sulfide minerals present in the blend from spectra measured during their combustion in a laboratory drop-tube setup, in which chemical reactions that take place in flash smelting furnaces can be reproduced. Controlled combustion experiments were conducted with two industrial concentrates and with high-grade mineral species as well, with a focus on pyrite and chalcopyrite. Exploratory analysis by means of Principal Component Analysis (PCA) applied on the spectral data depicted high correlation features among species with similar elemental compositions. Classification algorithms were tested on the spectral data, and a classification accuracy of 95.3% with a support vector machine (SVM) algorithm with a Gaussian kernel was achieved. The results obtained by the described procedures are shown to be very promising as a first step in the development of a predictive and analytical tool in search of fitting the current need for real-time control of pyrometallurgical processes.

Keywords: copper concentrate; pyrometallurgy; flash smelting; combustion; classification; spectroscopy; PCA; SIMCA; PLS-DA; k-NN; support vector machines

---

1. Introduction

The flash smelting process was developed in Finland in the late 1940s, and it has become one of the main copper production technologies in the world, given its high production and fast implementation capabilities at industrial and commercial scales. This process has attracted the interest of researchers for more than five decades, from the first works that allowed understanding the mineralogy and combustion kinetics of specific mineral particles, to modern works focused on the development and application of computer fluid dynamics (CFD) models [1–3].

Mineral oxidation at high temperatures is the core in such processes, since it involves complex energy and mass transfer mechanisms, as well as gaseous and intermediate species production.

The research related to this combustion process has tried to uncover the chemical and physical behaviors of those mineral particles present in copper concentrates in flash smelting conditions [4–6]. The species involved during the combustion absorb and emit energy with specific characteristics, which can be used to retrieve information about the process condition [7]. In particular, optical information at specific wavelengths has been used to describe the oxidation of the main sulfide minerals such as chalcopyrite ($CuFeS_2$) and pyrite (FeS_2) [4,6], to estimate ignition temperatures [8–10], single particle temperature [4,6], and to model particle size distributions [11].

In the combustion research field, visible and near-infrared (VIS-NIR) spectroscopic techniques are applied to characterize and retrieve relevant process information, e.g., Keyvan et al. [12,13] estimated natural gas flame temperature by means of the two-color pyrometry method; Romero et al. [14] developed a real-time temperature and composition monitoring system for natural gas flames in a glass production process; and Cai et al. [15] used least squares regression methods to fit coal flame spectral emission by means of using gray body models based on Planck's radiation law, with temperature and emissivity as regression parameters.

Recently, researchers have reported on spectral measurements from laboratory-scale experiments of copper concentrate combustion [16], and exploratory results depicted some spectral features related to the combustion process, such as Cu_xO spectral emissions at 606 nm and 616 nm, as well as a direct correlation among sulfur content in samples and broad band spectral intensity amplitudes. Furthermore, chemometrics techniques, i.e., analytical tools applied for the chemical characterization of samples, appear as a great tool to gain new insights from great volumes of data such as spectral data. One of these applications was shown in the work of Stumpe et. Al. [17]. The researchers achieved classifying different slag species coming from the steel industry by means of PCA and SIMCA (Soft Independent Modelling of Class Analogy) models applied to the mid-infrared spectra. Finally, industrial equipment applying radiometric techniques to control copper pyrometallurgical processes has been developed, the Optical Process Controller (OPC) monitoring system manufactured by Scandinavian Emissions Technology (Semtech) company. They discovered that lead species such as PbS and PbO can be used as tracers to follow the conversion process state [18,19].

In this work, classification and exploratory analysis algorithms are applied to characterize high-grade sulfide minerals by means of their VIS-NIR spectral radiation emitted from controlled combustion experiments. Spectral information is measured in absolute radiometric amplitudes with a previously-calibrated spectrometer and algorithms such as PCA, SIMCA, PLS-DA (Partial Least Squares-Discriminant Analysis), k-Nearest Neighbors (k-NN), and SVM are applied to visualize the spectral information behavior and to develop predictive classification models, so a short overview about these methods is introduced in the next section. Classification model performance and comparisons are evaluated during the training and validation process with cross-validation techniques, accuracy, and error metrics.

The work is organized as follows: in Section 2, a description of the combustion setup and an overview about the sensing techniques, preprocessing, and classification algorithms are given; in Section 3, the results are discussed, and in Section 4, some conclusions and future work are outlined.

2. Materials and Methods

In Figure 1, the experimental setup is depicted. This experimental setup was installed in the Metallurgical Engineering Department at the Universidad of Concepción, Concepción, Chile. The system mainly consisted of: (i) gas and solid feeding systems; (ii) a reaction zone heated by an electrical furnace; and (iii) an optical sensing system.

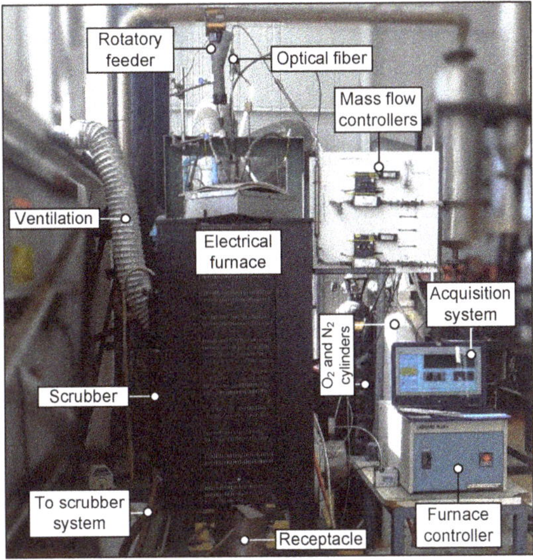

Figure 1. Experimental setup for combustion experiments.

The solid feeding system consisted of a LAMBDA DOSER® (Lambda CZ s.r.o., Brno, Czech Republic) of 0.2 L, with a feeding rate controller. Solids were fed by means of a water-cooled lance manufactured of stainless-steel; this system also refrigerates the optical fiber installed through the center to measure the radiation emitted in the zenithal position of the incandescent cloud of particles during combustion. The reaction zone was made with a stainless-steel tube of a 0.12-m inner diameter with a thickness of 3 mm, vertically positioned and heated on the surface by a controlled electrical furnace able to reach 1473 K. The furnace temperature was monitored by means of a K-type thermocouple. The process gas entering the reaction zone was a mixture of oxygen and nitrogen, and flows were controlled with mass flow controllers.

In Figure 2, the general data acquisition and preprocessing pipeline is depicted. Particles during combustion emitted radiation from the reaction zone, and the radiation was guided to a spectrometer by a cooled optical fiber probe (Avantes Inc., Louisville, CO, USA) specially designed for high-temperature environments. In this work, the VIS-NIR spectrometer USB4000 (Ocean Optis Inc., Dunedin, FL, USA) was used to acquire the spectral data in the range from 400–900 nm with an average spectral resolution of ~0.22 nm; also, the spectrometer was calibrated to measure the emitted radiation in absolute irradiance units (μW/(cm^2·nm)). The monitoring, acquisition, and spectrometer configuration stages were controlled with software developed in LabView™ (National Instruments Corporation, Austin, TX, USA). Moreover, from a spectral point of view, it was assumed that the main spectral features were emitted by particles in ignition and that hot gasses inside the reaction zone, e.g., SO$_2$, N$_2$, and O$_2$, were optically transparent in the analyzed spectral range.

In order to manipulate the spectral data, the applied algorithms assumed the data as a matrix $\mathbf{X}_{MxN}$ with columns representing the variables or sampling wavelengths, λ_i, $I = 1, \ldots , N$, with each row representing a spectrum, I_j, $j = 1, \ldots , M$, measured at some instant, as depicted in Figure 2. In this work, the number of acquired spectra for each experiment was related to material availability, and a total of $N = 2576$ discrete wavelengths in the 400–900 nm range were analyzed.

After the data were acquired, further preprocessing could be necessary to compensate for external perturbations such as particles size distribution and unstable feeding rates. Such methodologies are described next.

Signal processing and algorithm implementations were conducted in MATLAB® (The MathWorks, Inc., Natick, MA, USA) [20], with the PLS Toolbox 5.2 (Eigenvector Research, Inc., Manson, WA, USA) [21] and the Classification Learner App from the Machine Learning Toolbox™ (MATLAB®).

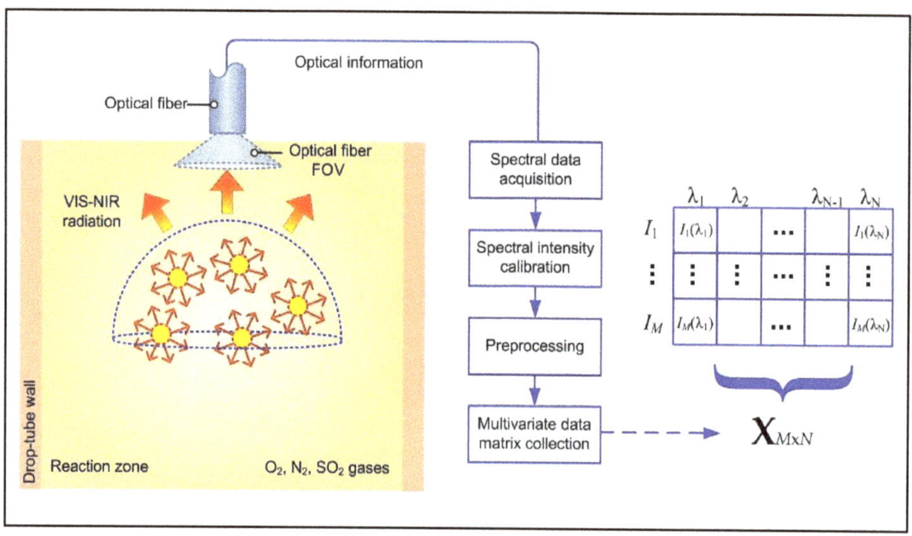

Figure 2. General spectral data acquisition and preprocessing pipeline.

2.1. VIS-NIR Spectral Signal Preprocessing

Spectral radiation emitted by objects at a high temperature are described by a continuous radiation spectral feature, $I_{bb}(\lambda,T)$, that follows a black body emission as a function of wavelength and object temperature. This radiation can be modeled by Planck's radiation law [16]. Since real bodies are not ideal emitters, an emissivity function, ε, that measures thermal energy emission efficiency was added to the model; thus, $I_c(\lambda,T) = \varepsilon \cdot I_{bb}(\lambda,T)$. This emissivity function can be wavelength independent (gray bodies) or wavelength dependent (real bodies). Moreover, in combustion processes, line, I_d, and molecular, I_m, emissions can be produced; thus, a measured spectrum can be modeled as $I(\lambda,T) = \varepsilon \cdot I_{bb}(\lambda,T) + I_d(\lambda) + I_m(\lambda) + n$, with n being a normally-distributed noise component.

As mentioned earlier, the acquired spectra require some preprocessing due to experimental issues, which produce high variance among spectral intensities at different acquisition times. In spectroscopy, some of the techniques suitable for external perturbation corrections are mainly divided into two groups: (i) transformation methods over spectral samples like MSC (Multiplicative Scatter Correction) and SNV (Standard Normal Variate) normalization methods, an; (ii) signal smoothing coupled with derivative procedures such as the Savitzky–Golay (SG) algorithm. An exhaustive description of the aforementioned methods can be found in [22]. It has been shown in works by some authors that applying preprocessing algorithms improves the performance in regression or classification models developed from the data [23].

2.2. Principal Component Analysis

After the preprocessing stage, an exploratory analysis was performed. For this purpose, PCA [24] was implemented with MATLAB® and the PLS Toolbox. The goal of this method was to approximate the data matrix $\mathbf{X}_{MxN}$ by the product of two matrices:

$$\mathbf{X} = \mathbf{T} \cdot \mathbf{L}^T \tag{1}$$

where $\mathbf{T}$ is the score matrix with M rows and d columns equal to the number of Principal Components (PCs), and the $\mathbf{L}$ matrix is the loading matrix with d columns and N rows. This analysis allows reducing the dimensionality of the original data to visualize their behavior easily in a reduced space; it also allows assessing the most important variables that contribute to the variance in the original dataset. This exploratory method is also the base for classification algorithms such as the SIMCA and PLS-DA methods.

2.3. Classification Methods

In this work, the k-NN, SIMCA, PLSDA, and SVM classification methods were implemented. To implement such methods, a spectral training set was needed, with each spectrum representing a known class or category, e.g., 0 and 1 for chalcopyrite and pyrite, respectively, then, by having this set or by developing the classification model, predictions on new spectral samples can be performed. Figure 3 summarizes a general implementation of the classification algorithms to conduct predictions for new data.

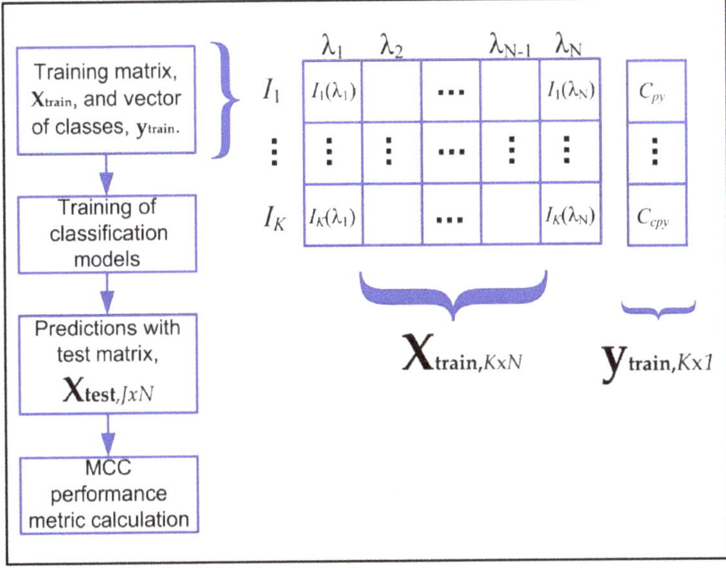

Figure 3. General classifiers training and prediction pipeline.

The k-NN method classifies an unknown spectrum by taking a distance measure (Euclidean or Mahalanobis distance) to its nearest neighbors, of known categories. Therefore, an unknown sample is classified according to the classes of its closest neighbors [25]. The SIMCA method [26] calculates a PCA model on each spectral training dataset belonging to a known class. Then, it defines boundaries around the reduced sample space for each class with a given probability, commonly of 95%, which allows classes to overlap and, thus, a sample to belong to one or more categories with a defined probability.

The PLS-DA method is an adaptation of the Partial Least Squares regression method (PLS). In this method, the target class or dependent variables are required; thus, a matrix Y is generated containing the encoded classes as 0 and 1. PLS-DA reduces the dimensionality of measured variables, but in this case, through partial least squares. Once the new Latent Variables (LVs) are calculated, the discriminant analysis is carried out, and the boundaries between the classes are established. The classification of new samples in the discriminant analysis is based on their probability to belong to one or another class: the class with higher probability is assigned to the sample [27]. Finally, the SVM method constructs linear decision surfaces over the original input vectors (samples) or mapped vectors into a high-dimension feature space through the implementation of kernel functions; in this work, two different types of kernel functions were investigated: linear and Gaussian (RBF) [28].

For the accuracy of the trained model, confusion matrices (also known as misclassification matrices) are presented [24]. These tables summarize the classification performance by depicting the number of: False Positive estimations (FP); False Negative estimations (FN); True Positive estimations (TP) and True Negative estimations (TN). This allows a more detailed analysis than the mere proportion of correct classifications (accuracy). Then, Matthews's correlation coefficient (MCC) is also estimated; this metric can be estimated from the confusion matrix as:

$$\text{MCC} = \frac{\text{TP} \times \text{TN} - \text{FP} \times \text{FN}}{\sqrt{(\text{TP} + \text{FP})(\text{TP} + \text{FN})(\text{TN} + \text{FP})(\text{TN} + \text{FN})}} \tag{2}$$

For the defined metric, Equation (2), MCC takes values between −1 and +1: MCC = +1 represents a perfect prediction; MCC = −1 indicates a total disagreement between the predicted and observed classes, and value of MCC = 0 represents a prediction no better than a random prediction. Note that this metric can be only used in binary classification problems, in our case, we were only predicting the presence of chalcopyrite or pyrite, since they are the main mineralogical species present in the copper concentrates analyzed in this work.

2.4. Raw Materials and Experimental Design

Raw materials used in this work consisted of high-grade sulfide minerals mainly present in copper concentrates. Table 1 depicts these species together with their p80 size parameter, chemical formula, and abbreviations. Some of them were acquired from Ward's Natural Science and Northern Geological Suppliers, others by means of local suppliers.

Table 1. List of sample minerals.

Minerals	Abbreviations	Formula	p80
Chalcopyrite	Cpy	$CuFeS_2$	37 μm
Pyrite	Py	FeS_2	35 μm
Bornite	Bn	Cu_5FeS_4	32 μm
Covelline	Cv	CuS	32 μm
Chalcocite	Cs	Cu_2S	37 μm
Pyrrhotite	Po	FeS	35 μm
Enargite	Enr	Cu_3AsS_4	42 μm

Minerals were prepared with standard laboratory procedures to achieve dry and similar size distributions. The mineralogy and size distribution of the samples were determined by means of X-Ray Diffraction (XRD) and laser diffraction, respectively. Table 2 shows the qualitative analysis produced by the XRD method.

Table 2. Qualitative results by X-ray diffraction.

Sample	Mineralogical Composition						
	Cpy	Py	Bn	Cs	Cv	Po	Enr
Chalcopyrite	***	Tr	-	*	-	-	-
Pyrite	-	***	-	-	-	-	-
Bornite	*	-	***	-	-	-	-
Chalcocite	*	-	-	***	-	-	-
Covelline	Tr	-	-	*	***	-	-
Pyrrhotite	-	*	-	-	-	***	-
Enargite	Tr?	Tr	-	-	-	-	***

*** = most abundant phase, ** = abundant phase, * = minority phase, - = no presence, Tr = trace phase, ? = uncertainty phase.

Two different copper concentrates were also used as raw material, and their mineralogical composition is shown in Table 3.

Table 3. Mineralogical composition of copper concentrate measured with a QEMSCAN®.

Mineral (%)	Concentrate A	Concentrate B
Chalcopyrite	32.7	66.7
Pyrite	45.5	16.5
Bornite	3.2	2.1
Covelline	0.8	0.1
Quartz	0.9	1.7
Muscovite	1.3	2.0
Others	15.6	10.9

The experimental design considered the combustion of mineral samples under fixed operating conditions for all experiments, and such conditions were assessed as optimal from exploratory experiments to ensure high signal to noise ratios, while the values were: a furnace operating temperature of 1273 K and an 80%v O_2, 20%v N_2 process gas. Nitrogen and oxygen flows were adjusted accordingly to ensure laminar flow conditions inside the drop-tube. For each experiment, 0.03 kg of sample were fed to the drop-tube.

Finally, combustion products were collected by means of a receptacle located at the bottom of the drop-tube; the receptacle was water cooled to stop as fast as possible the chemical transformations in order to have representative samples from the high-temperature oxidation process., and the products were treated to conduct further analysis by QEMSCAN® (Quantitative Evaluation of Minerals by SCANning electron microscopy, FEI Company, Hillsboro - Oregon, USA) technology.

3. Results and Discussion

In Figure 4, average measurements of calibrated spectra from each species are depicted, and differences among spectral intensities along the sensed spectral range are observed, with pyrite emission producing the highest intensity pattern. Moreover, the pyrite spectrum shows a pronounced peak at 588 nm and a doublet at 765.8–769.3 nm, and the same peaks appear in the pyrrhotite, but with less intensity. These signals were associated with sodium and potassium emissions, respectively, in previous works [16]. On the other hand, the chalcopyrite spectrum shows slightly perceptible peaks, while the other mineral species show only continuum spectral patterns.

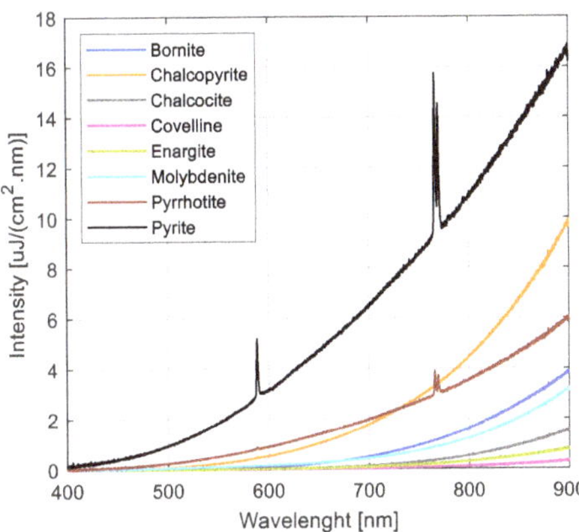

Figure 4. Average emission spectra recorded during the combustion of seven different minerals.

In order to compare the emission spectra from the different mineral species, a single measurement matrix was constructed containing all the data, following the structure described in Figure 2. Because the size of the measurement matrix was very large, the application of principal component analysis (PCA) was chosen as an alternative to visualize the patterns of possible mineral species. Results from PCA application are depicted in the scatterplot of Figure 5. The score analysis shows four very marked groups, two related to chalcopyrite and pyrite emissions, one produced by pyrrhotite scores, slightly overlapping on the pyrite spectra, and one group with the other species, which indicates a high correlation among their spectral emission patterns.

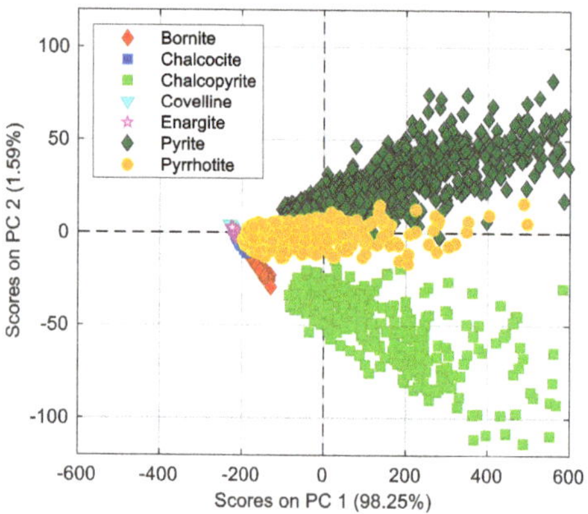

Figure 5. PC1–PC2 scores from PCA applied to the mean centered data matrix.

The pyrrhotite scores' behavior can be explained by the fact that at temperatures above 873 K, the oxidation of chalcopyrite and pyrite occurs mainly through the decomposition of sulfur to produce FeS according to the reactions:

$$CuFeS_{2(s)} + 1/2O_{2(g)} \rightarrow 1/2Cu_2S_{(s)} + FeS_{(s)} + 1/2SO_{2(g)} \tag{3}$$

$$FeS_{2(s)} + O_{2(g)} \rightarrow FeS_{(s)} + SO_{2(g)} \tag{4}$$

The spectra from the combustion of the sulfides presented in the previous equations can be confused with the spectra of the sulfide mineral species. As mentioned earlier, the overlapping among the score groups produced by bornite, chalcocite, covelline, and enargite [29–31] emissions can be justified by the fact that the product of their thermal decomposition is Cu_2S (reactions thermodynamically favorable under the temperature attained by the combustion flames in each experiment), as shown in the following reactions:

$$Cu_5FeS_4 \rightarrow 5/2Cu_2S + FeS + 1/4S_{2(g)}, \tag{5}$$

$$Cu_3AsS_4 \rightarrow 3/2Cu_2S + 1/2\,As_2S_3 + 1/2S_{2(g)}, \tag{6}$$

$$CuS \rightarrow 1/2Cu_2S + 1/4S_{2(g)}, \tag{7}$$

Moreover, the results of PCA applied on copper sulfide spectra had a very marked separation with the chalcopyrite scores; in this case, this separation was given by the PC1, with most of chalcopyrite scores located on the negative side of PC2.

In Table 4, the mineralogical composition of the combustion products is summarized. It can be seen that products from enargite and covelline combustion have a high content of the Cu_2S phase (chalcocite). On the other hand, bornite and chalcocite partially reacted, and they also depicted low intensity profiles, so their radiation was prone to be overshadowed by the radiation emitted by the furnace walls.

Table 4. Mineralogical composition of sulfide combustion products determined by QEMSCAN®.

Products (%)	Py	Po	Cpy	Enr	Bn	Cv	Cs
Delafossite	0.04	0.02	33.70	0.70	0.62	0.24	0.36
FeOx	82.25	94.33	22.90	1.24	3.34	0.51	0.55
Chalcocite	0.00	0.00	2.52	52.01	19.38	77.38	83.27
Cuprite	0.00	0.00	1.43	28.92	18.44	16.63	15.68
Bornite	0.03	0.00	7.80	0.18	44.53	0.33	0.02
Enargite	0.00	0.00	0.00	15.41	0.25	0.00	0.00
Covellite	0.00	0.00	0.01	0.10	8.84	3.13	0.05
Other	17.68	5.65	31.64	1.44	4.60	1.78	0.07

In Figure 6a, the plot of PCs loadings depicts the variables' (wavelengths) behavior for pyrite and chalcopyrite combustion spectra. The peaks mentioned above and less intense peaks at 779.1 and 793.9 nm are observed. These peaks have been previously reported and may be associated with iron species [16]. These peaks are also observed in the loadings obtained with PCA using only pyrite spectra. For chalcopyrite spectra, loadings depict peaks at 606 and 616 nm (Figure 6b), and they are associated with the presence of copper oxides [32]. Note that loading vectors only give an idea of the wavelengths' contribution to spectral data variance and that in any case, their amplitude can be interpreted as a relative concentration in the original spectra; however, they can be seen as a first approach to elucidate the structure of spectral patterns hidden in the data because of their weak emissions. Loading analysis for other species presented no relevant spectral patterns, so they are not shown in this work.

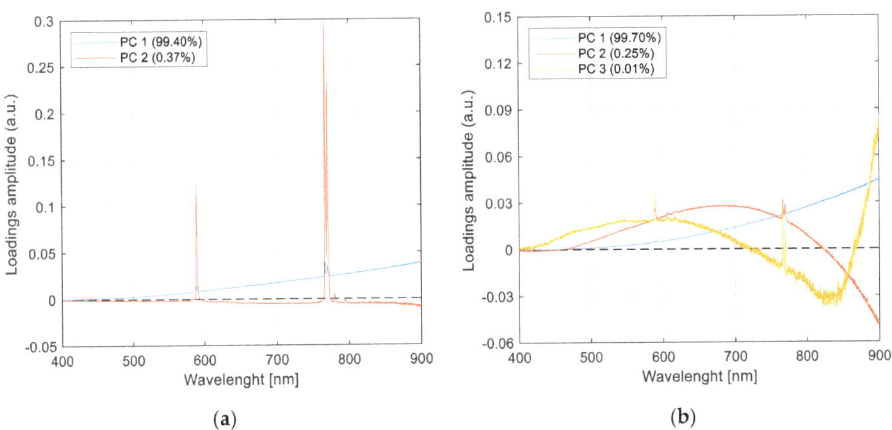

Figure 6. (a) PC1,PC2 loadings plots of pyrite combustion and (b) PC1–PC3 loadings plots of chalcopyrite combustion.

Due to the promising results obtained with PCA, supervised classification methods such as k-NN, PLS-DA, SIMCA, and SVM were applied. In this section, only chalcopyrite and pyrite emissions are considered for analysis. To accomplish this, a training matrix was constructed from the emission spectra that presented the best differentiation between mineral species, as depicted in Figure 3. Under this assumption, 750 chalcopyrite and 750 pyrite spectra were chosen randomly. Finally, trained classification models were evaluated on 500 spectra from chalcopyrite and pyrite combustion, spectra from Copper Concentrate A and Copper Concentrate B combustion, and finally, spectra from mixtures of pyrite/chalcopyrite in proportions of 30% Cpy-70% Py and 70% Cpy-30% Py combustion.

From the set of applied preprocessing methods, the mean centering approach was chosen since it presented a good justification of the accumulated variance from exploratory PCA analysis with a good segregation of sample scores and low values of the root mean squared error of cross-validation for the different methods' implementation. In this work, a 10-fold cross-validation method was implemented to estimate the optimum parameters of the trained models. Table 5 summarizes the optimum assessed parameters for each implemented method.

Table 5. Calibration of the models depicting optimum parameters and performances evaluated with a 10-fold cross-validation procedure.

Model	Optimum Parameters	MCC
k-NN	2 neighbors	1
SIMCA	2 PCs for each species	0.983
PLS-DA	7 LVs	1
SVM	Fine Gaussian kernel	1

Table 6 summarizes the results of predictions over the test matrices by using the optimal models, the values of the MCC metric, and confusion matrices. It can be seen that during the predictions, the k-NN model was not considered appropriate for the detection of pyrite combustion spectra, because it had a low specificity (a large number of false positive samples); the same issue is observed with SIMCA, with higher rates of false positives for both species; the PLS-DA and SVM methods show the best classification results for the sulfide mineral species' predictions. In the case of predicting the class of copper concentrates, Concentrate A was mainly classified with a higher presence of pyrite, and the opposite can be observed for Concentrate B, which was accurate by considering their mineralogical composition; see Table 3. The same results can be observed from the binary mixture combustion, and the algorithms predicted the high presence of pyrite or chalcopyrite species, accordingly.

Table 6. Prediction of chalcopyrite, pyrite, concentrates, and sulfide mixture by the trained models.

Model		Model Prediction					
		Cpy	Py	ConA	ConB	Py 70%-Cpy 30%	Py 30%-Cpy 70%
k-NN	MCC	0.886		-			-
	Cpy	227	6	250	250	250	240
	Py	23	244	0	0	0	10
SIMCA	MCC	0.721		-			-
	Cpy	240	64	0	241	207	84
	Py	40	186	250	9	43	166
PLS-DA	MCC	0.900		-			-
	Cpy	236	11	2	244	235	160
	Py	14	239	248	6	15	90
SVM	MCC	0.953		-			-
	Cpy	250	12	6	250	249	235
	Py	0	238	244	0	1	15

4. Conclusions

From the results, it can be concluded that depending on the degree of reaction of sulfide species, the spectra emitted can show patterns that allowed them to be differentiated, such as the pyrite and pyrrhotite spectra in which emission peaks can be observed at 588, 765.8, and 769.3 nm, while species like chalcopyrite required a multivariate analysis to uncover these peaks. In this case, by applying PCA to the spectral datasets, peaks related to copper phases (606 and 616 nm) and others related to the oxidation of iron sulfides (779.1 and 793.9 nm) were found. These results allowed evaluating the efficiency of the classification models by means of methods such as k-NN, SIMCA, PLS-DA, and SVM. With all these methods, a good degree of prediction was observed against pyrite and chalcopyrite spectra, while applying these methods to the spectra of copper concentrates' combustion or binary mixtures, the results were accurate in the sense that a higher presence of the two analyzed species was predicted. Finally, the classification results with an SVM approach and with a Gaussian mapping function of the original spectra generated the best classification results with 95.3% accuracy. In future work, we will extend this analysis to perform regression predictions so that an estimation of the proportion of sulfide mineral species during combustion in real scenarios can be performed.

Author Contributions: Conceptualization, W.D. and C.T.; methodology, W.D., G.R., C.T., and V.P.; software, W.D.; validation, W.D., C.T., and V.P.; formal analysis, W.D., C.T., and P.C.; investigation, W.D.; resources, E.B., R.P.; data curation, C.T. and V.P.; writing, original draft preparation, W.D.; writing, review and editing, C.T., V.P., E.B., W.D., P.C., and R.P.; visualization, W.D.; supervision, C.T.; project administration, E.B.; funding acquisition, R.B. and E.B.

Funding: This research was funded by the CONICYT, Anillo Minería ACM170008, and by Fondef IT under Grant Number 16M10029. The work of C. Toro was supported by the CONICYT Fondecyt/Postdoctorate grant under Project Number 3170897. The work of Pablo Coelho was supported by CONICYT PAI/CONVOCATORIA NACIONAL SUBVENCIÓN A LA INSTALACIÓN EN LA ACADEMIA, 2018 (77180078).

Acknowledgments: We thank the Metallurgical Engineering Department at the University of Concepción for giving access to their facilities, allowing us to conduct the experiments reported in this work.

Conflicts of Interest: The authors declare no conflict of interest. The funders had no role in the design of the study; in the collection, analyses, or interpretation of data; in the writing of the manuscript; n\or in the decision to publish the results.

References

1. Jorgensen, F.R.A.; Segnit, E.R. Mineralogy of the products of the flash smelting of chalcopyrite. In Proceedings of the 25th International Geological Congress, Sidney, Australia, 16–25 August 1976; pp. 575–576.

2. Jorgensen, F.R.A.; Segnit, E.R. Copper flash smelting simulation experiments. *Proc. Australas. Inst. Min. Metall.* **1977**, *261*, 39–46.

3. White, M.; Haywood, R.; Ranasinghe, D.J.; Chen, S. The development and application of a cfd model of copper flash smelting. In Proceedings of the Eleventh International Conference on CFD in the Minerals and Process Industries, Melbourne, Australia, 7–9 December 2015.

4. Jorgensen, F.R.A. Combustion of pyrite concentrate under simulated flash-smelting conditions. *Trans. Inst. Min. Metall. Section C* **1981**, *90*, C1–C9.

5. Jorgensen, F.R.A. Heat transfer mechanism in ignition of nickel sulphide concentrate under simulated flash smelting conditions. *Proc. Australas. Inst. Min. Metall.* **1979**, *271*, 21–25.

6. Sohn, H.Y.; Chaubal, P.C. The ignition and combustion of chalcopyrite concentrate particles under suspension-smelting conditions. *Metall. Trans. B* **1993**, *24*, 975–985. [CrossRef]

7. Howell, J.R.; Menguc, M.P.; RSiegel, R. *Thermal Radiation Heat Transfer*, 6th ed.; CRC Press: Boca Raton, FL, USA, 2015; pp. 441–488.

8. Jorgensen, F.R.A.; Zuiderwyk, M. Two-colour pyrometer measurement of the temperature of individual combusting particles. *J. Phys. E Sci. Instrum.* **1985**, *18*, 486–491. [CrossRef]

9. Morgan, G.J.; Brimacombe, J.K. Kinetics of the flash converting of MK (chalcocite) concentrate. *Metall. Mater. Trans. B* **1996**, *27*, 163–175. [CrossRef]

10. Wilkomirsky, I.; Otero, A.; Balladares, E. Kinetics and reaction mechanisms of high-temperature flash oxidation of molybdenite. *Metall. Mater. Trans. B* **2010**, *41*, 63–73. [CrossRef]

11. Laurila, T.; Hernberg, R.; Oikari, R.T.; Mikkola, P.; Ranki-Kilpinen, T.; Taskinen, P. Pyrometric temperature and size measurements of chalcopyrite particles during flash oxidation in a laminar flow reactor. *Metall. Mater. Trans. B* **2005**, *36*, 201–208. [CrossRef]

12. Keyvan, S.; Rossow, R.; Romero, C. Blackbody-based calibration for temperature calculations in the visible and near-IR spectral ranges using a spectrometer. *Fuel* **2006**, *85*, 796–802. [CrossRef]

13. Keyvan, S.; Rossow, R.; Romero, C.; Li, X. Comparison between visible and near-IR flame spectra from natural gas-fired furnace for blackbody temperature measurements. *Fuel* **2004**, *83*, 1175–1181. [CrossRef]

14. Romero, C.; Li, X.S.; Keyvan, S.; Rossow, R. Spectrometer-based combustion monitoring for flame stoichiometry and temperature control. *Appl. Thermal Eng.* **2005**, *25*, 659–676. [CrossRef]

15. Cai, X.; Cheng, Z.; Wang, S. Flame measurement and combustion diagnoses with spectrum analysis. *AIP Conf. Proc.* **2007**, *914*, 60–66.

16. Arias, L.; Torres, S.; Toro, C.; Balladares, E.; Parra, R.; Loeza, C.; Villagrán, C.; Coelho, P. Flash smelting copper concentrates spectral emission measurements. *Sensors* **2018**, *18*, 2009. [CrossRef] [PubMed]

17. Stumpe, B.; Engel, T.; Steinweg, B.; Marschner, B. Application of PCA and SIMCA statistical analysis of FT-IR spectra for the classification and identification of different slag types with environmental origin. *Environ. Sci. Technol.* **2012**, *46*, 3964–3972. [CrossRef] [PubMed]

18. Persson, W.; Wendt, W.; Demetrio, S.J. Use of optical on-line production control in copper smelters. In Proceedings of the 4th International Conference COPPER 99-COBRE 99, Phoenix, AZ, USA, 10–13 October 1999; pp. 491–503.

19. Persson, W.; Wendt, W. Optical spectroscopy for process monitoring and production control in ferrous and non-ferrous industry. In *Modeling, Control, and Optimization in Ferrous and Non-Ferrous Industry, Proceedings of the Materials Science & Technology 2003 Conference, Chicago, IL, USA, 9–12 November 2003*; The Minerals, Metals & Materials Society: Pittsburgh, PA, USA, 2003; pp. 177–191.

20. MATLAB. Available online: https://la.mathworks.com/products/matlab.html (accessed on 21 July 2019).

21. PLS_Toolbox, Eigenvector. Available online: https://eigenvector.com/software/pls-toolbox/ (accessed on 21 July 2019).

22. Rinnan, A.; van den Berg, F.; Engelsen, S.B. Review of the most common pre-processing techniques for near-infrared spectra. *TrAC Trends Anal. Chem.* **2009**, *28*, 1201–1222. [CrossRef]

23. Parente, A.; Sutherland, J.C. Principal component analysis of turbulent combustion data: Data pre-processing and manifold sensitivity. *Combust. Flame* **2013**, *160*, 340–350. [CrossRef]

24. Lindon, J.C.; Tranter, G.E.D.; Koppenaal, D. *Encyclopedia of Spectroscopy and Spectrometry*, 2nd ed.; Academic Press: Oxford, UK, 2016; pp. 1704–1709.

25. Otto, M. *Chemometrics: Statistics and Computer Application in Analytical Chemistry*, 3rd. ed.; John Wiley & Sons: Weinheim, Germany, 2016; pp. 135–211.

26. Ballabio, D.; Grisoni, F.; Todeschini, R. Multivariate comparison of classification performance measures. *Chem. Intell. Lab. Sys.* **2018**, *174*, 33–44. [CrossRef]

27. Chevallier, S.; Bertrand, D.; Kohler, A.; Courcoux, P. Application of PLS-DA in multivariate image analysis. *J. Chemometr.* **2006**, *20*, 221–229. [CrossRef]

28. Cortes, C.; Vapnik, V. Support-vector networks. *Mach. Learn.* **1995**, *5*, 273–297. [CrossRef]

29. Chaubal, P.C.; Sohn, H.Y. Intrinsic kinetics of the oxidation of chalcopyrite particles under isothermal and nonisothermal conditions. *Metall. Mater. Trans. B* **1986**, *17*, 51–60. [CrossRef]

30. Padilla, R.; Fan, Y.; Wilkomirsky, I. Decomposition of enargite in nitrogen atmosphere. *Can. Metall. Q.* **2001**, *40*, 335–342. [CrossRef]

31. Dunn, J.G.; Muzenda, C. Thermal oxidation of covellite (CuS). *Thermochim. Acta* **2001**, *369*, 117–123. [CrossRef]

32. Gole, J.L. Oxidation of small metal and metalloid molecules. In *Gas-Phase Metal Reactions*, 1st ed.; Fontjin, A., Ed.; Elsevier: New York, NY, USA, 1992; pp. 596–597.

Article

Deposits in Gas-fired Rotary Kiln for Limonite Magnetization-Reduction Roasting: Characteristics and Formation Mechanism

Xianghui Fu [1,2,*], Zezong Chen [2], Xiangyang Xu [3], Lihua He [1] and Yunfeng Song [1,*]

[1]   School of Metallurgy and Environment, Central South University, Changsha 410083, China
[2]   Changsha Research Institute of Mining and Metallurgy Co. Ltd., Changsha 410012, China
[3]   School of Minerals Processing and Bioengineering, Central South University, Changsha 410083, China
*   Correspondence: fuxh@csu.edu.cn (X.F.); syfeng@csu.edu.cn (Y.S.); Tel: +86-138-7318-3635 (X.F.); +86-151-1627-8929 (Y.S.)

Received: 4 June 2019; Accepted: 5 July 2019; Published: 8 July 2019

Abstract: The formation mechanism of deposits in commercial gas-fired magnetization-reduction roasting rotary kiln was studied. The deposits ring adhered on the kiln wall based on the bonding of low melting point eutectic liquid phase, and the deposit adhered on the air duct head by impact deposition. The chemical composition and microstructure of the deposits sampled at different locations varied slightly. Besides a small amount of quartz and limonite, main phases in the deposits are fayalite, glass phase and magnetite. The formation of the deposits can be attributed to the derivation of low melting point eutectic of fine limonite and coal ash, and the solid state reaction between them. Coal ash, originated from the reduction coal, combining together with fine limonite particles, results in the accumulation of deposits on the kiln wall and air duct. Fayalite, the binder phase, was a key factor for deposit formation. The residual carbon in limonite may cause an over-reduction of limonite and produce FeO. Amid the roasting process, SiO_2, originated from limonite and coal ash, may combine with FeO and reduce the liquefaction temperature, therewith liquid phase generates at high temperature zone, which can significantly promote the growth of deposits.

Keywords: limonite; magnetization reduction roasting; rotary kiln; deposit; fayalite; FeO; liquid phase

---

1. Introduction

Due to the depletion of easy-processing iron ores, much attention has been paid to the utilization of refractory ores in recent years [1–5]. Limonite ore is a common refractory iron oxide ore with low grade, weak magnetic property and high gangue content. Although limonite ore is poorly responsive to conventional beneficiation techniques, magnetizing roasting-magnetic separation approach has been proved a promising solution for such refractory iron oxide ore beneficiation [1,6–9]. The limonite ore becomes porous, easy-to-be-reduced and the iron grade can be naturally enriched by the removal of crystal water which was banded with iron oxide after roasting process, then qualified iron concentrate can be obtained by the sequential magnetic separation [10].

Industrial-scale magnetization reduction roasting (MRR) was first implemented in Ukraine in the 1970s [11]. The MRR technology mainly include shaft furnace MRR, rotary kiln MRR, fluidized MRR, etc. [12]. Generally, shaft furnace MRR technology has problems such as low iron recovery rate, inefficient use of fine ores and high energy consumption [13]. Fluidized MRR technology has been reported by many researchers [14–17], but due to the equipment design and control problems in its industrialization, it stays still at laboratory or pilot level. Rotary kiln magnetization reduction roasting technique has been applied in limonite processing lines. To explore the limonite ore resource in Xinjiang, China, a Φ 4 m × 60 m rotary kiln MRR system was designed [18]. Till late of 2010, five MRR systems with the single capacity of 55 t/h were constructed and stable operation was achieved. With this roasting-magnetic separation process, the average concentrate grade reached 61–62%, and the iron recovery amounted to about 90%, which realized the economic and effective utilization of limonite ore in actual practice.

However, amid the MRR production, it was usually found that there were deposits adhered on the kiln wall at the discharge end and the top of the air duct in the kiln [18]. The deposit formation is a serious problem which can significantly reduce the roasting product quality and the operation efficiency. In grate-kiln for hematite pellet production, the characteristics and mechanism of the deposits, especially, the influence of coal ash, were widely studied. The results show that the main formation mechanism of deposits in rotary kiln is the crystallization and diffusion hematite. The liquid phase plays a secondary role in the deposit formation [19,20], while the silicate phase also plays an important role in deposit solidification [21]. Coal ash can enhance the bonding strength of the iron ore powder compact [22]. A combination of coal ash, silicide glass phase, alum inosilicate molten phase and low melting point liquid phase promotes the formation and accumulation of deposits, which eventually results in deteriorate normal production conditions. In addition, studies show the formation of deposit mainly depends on the accumulation of powder stocked on kiln wall, and the powder concretion speed turns faster at higher temperature [23]. Research has shown that [24] compared to other fuels, the usage of clean fuel is greatly beneficial in reducing of deposit formation rate.

As the commercial application of limonite MRR production is rare, there were few reports on its deposit phenomena. Run-of-mine limonite ore was processed by the aforementioned Φ 4 m × 60 m rotary kiln MRR plant in Xinjiang. Natural gas is used as the main fuel, while the reduction coal, mixed with limonite ore as raw material, and a certain amount of pulverized coal was injected into the rotary kiln by a coal gun to adjust the atmosphere and temperature. The total amount of coal was about 4–8% (mass weight) of limonite ore. Serious deposit problem occurred during its beginning operation. Therefore, it is essential to investigate the deposit formation mechanism. As the MRR and oxidizing pellet processes are different in terms of raw materials, reaction atmosphere and temperature, the formation mechanism and characteristics of the deposits may also be different. The formation mechanism of deposit in a gas-fired rotary kiln for limonite MRR is still not very clear, so it is essential to reveal the mechanism and take some practical measures to mitigate these occurrences to improve product quality and productivity.

2. Materials and Methods

2.1. Materials

2.1.1. Chemical Analysis on Raw Materials

The chemical compositions and ignition loss of limonite ore are listed in Table 1. The coal properties and ash components are listed in Tables 2 and 3.

Table 1. Chemical Compositions and Ignition Loss of Limonite Ore (wt%).

Total Fe	FeO	Fe_2O_3	SiO_2	Al_2O_3	CaO	MgO	MnO	K_2O	Na_2O	S	Ig
35.85	10.2	39.92	18.89	3.79	2.22	2.42	1.86	1.02	0.092	0.26	15.8

Table 2. Property of the Coal (wt%).

Moisture/%	Ash/%	Volatile Matter/%	Fixed Carbon/%	Qgr/(MJ·Kg^{-1})	Coking Properties (1–8)
2.3	10.28	30.02	57.32	28.63	2

Table 3. Chemical Compositions of Coal Ash (wt%).

SiO$_2$	Fe$_2$O$_3$	Al$_2$O$_3$	CaO	MgO	K$_2$O	Na$_2$O
55.42	9.81	20.18	2.68	2.8	3.56	0.42

2.1.2. Locations in the Kiln Where Deposits Were Collected

Three kinds of deposit samples from different locations, one on the air duct and the other two on the wall, were studied. The collection position on the cross section of the deposit ring on kiln wall is shown in Figure 1. The samples collected from the upper layer, i.e., above the midline of the longitudinal section of the deposit ring were labeled as S1, while those collected beneath the middle line adjacent to the kiln wall were labeled as S2. Four samples were collected randomly of each layer along the kiln wall deposit ring circumference. The size of a single sample was about 20 × 20 × 15 cm (length × width × thickness). The deposit on the top of the first of the four air duct near the discharge end was marked as S3, and there were no deposits on other air ducts behind.

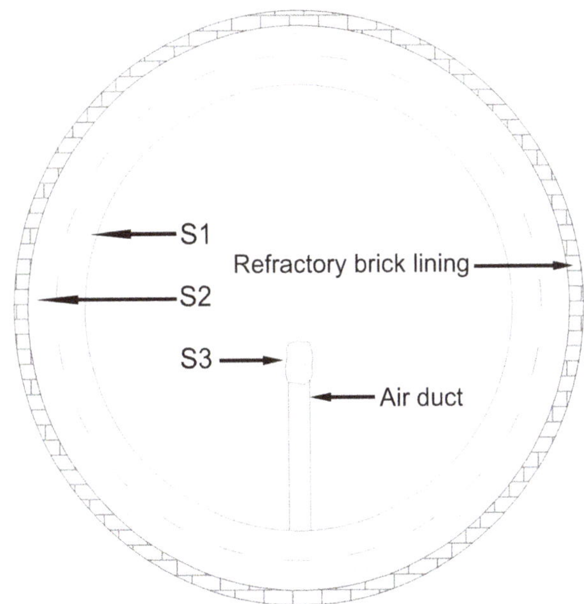

Figure 1. Schematic diagram of sampling locations of different deposits.

2.2. Characterization Methods

The chemical compositions of limonite ore and deposit samples were determined using Chinese Iron Ore Chemical Analysis Methods Standard [25]. The chemical composition of coal ash and the analysis of steam coal were conducted according to GB/T 1574-2007and GB/T 212-2008, respectively. The soft-melting characteristics of coal ash, limonite ore and deposit samples were analyzed according to GB/T 212-2008. The phase composition of deposit samples was determined by X-ray diffraction technique (XRD, Rigaku D/max-rA, Rigaku Co., Akishima, Japan). A scanning electron microscopy (SEM, JSM-6490LV, JEOL Co., Ltd, Akishima, Japan) coupled with energy dispersive spectra analyzer (EDS, NEPTUNE TEXS HP, EDAX Inc. Co., Ltd., McKee Drive Mahwah, NJ, USA) was used to investigate the elemental distributions of the deposit samples. The mineralogy identification analysis of the deposit samples was observed using an automatic polarizing microscope (DM4500P, Leica Microsystems, Wetzlar, Germany). FactSage® 7.2 software was used to calculate the liquid phase range and liquefaction temperature change of the iron silicate phase in the deposit samples at different FeO contents in Al_2O_3-SiO_2-CaO-FeO system with the corresponding databases "FToxid" and "FactPS".

3. Results and Discussion

3.1. In-Line Observation of Deposits

The deposit ring of limonite MRR production generates at the discharge end. Its height increased gradually along with the extension distance to the charge end. Located at a distance of 8 to 10 m in the kiln from the discharge end, the deposit ring can accumulate to over one-meter high. Its height drops rapidly after passing through the high temperature zone. Figure 2 shows the actual situation and longitudinal section of the deposit ring in the rotary kiln. The accumulative deposit ring height may affect the flow of raw materials and flue gas. The temperature in the region before the ring increased, while that behind the ring declined remarkably, resulting in deterioration of process conditions. When the height of the deposit ring attained 0.8–1.0 m or more, as illustrated in Figure 2b, the flow of flue gas and raw materials will be hindered by the narrowed channel of the deposition ring. With the accumulation of a large amount of raw materials between the deposit ring and the charge end of the kiln, the motor risks overloading. If the deposit ring cannot fall off automatically, it will artificially interfere to pause the operation and peel off the deposit ring after cooling down. As the deposit ring needs 3 to 5 days to be removed per time, this reduces MRR production efficiency.

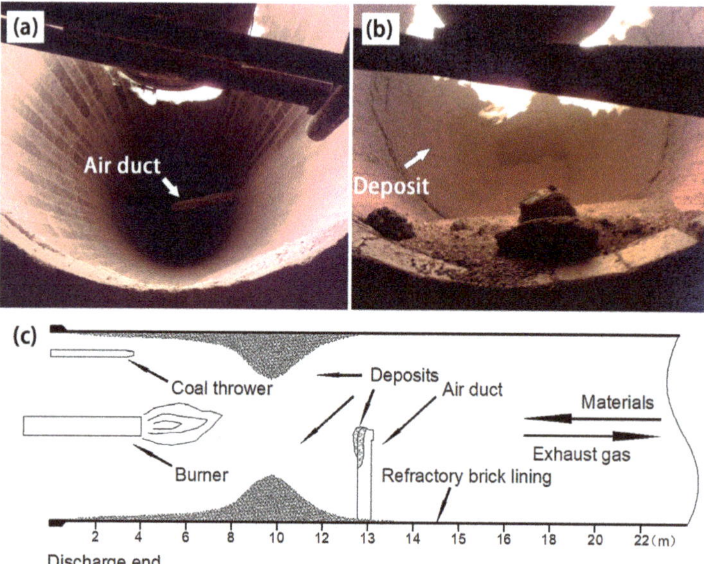

Figure 2. Digital photos of internal view of the Φ 4 m × 60 m limonite magnetization reduction roasting (MRR) rotary kiln: (**a**) without deposit; (**b**) with serious deposits; (**c**) cross section schematic of the kiln.

Compared with the oxidized pellets production, which needs to be roasted generally at temperatures over 1200 °C [20], the working temperature for the limonite MRR production is only 700–800 °C. Nevertheless, serious deposition occurs and the deposits grow very fast. The height of the deposit ring on kiln wall can reach 0.4–0.8 m in 5–10 days and has seriously negative impact on limonite MRR production.

3.2. Appearance of Deposits

Figure 3 shows the appearance of the deposit samples. It can be seen that all these samples are of grayish-black, while some of the deposits are stained with red Fe_2O_3 powder because the atmosphere in the kiln was turned into oxidizing when quitting production. Showing fused state and metallic luster, sample S1 possesses well-developed pores, and there is a lot granular limonite adhered to its surface (Figure 3a). Sample S2 is mainly composed of fine-grained materials with obvious stratification, which contains a small amount of granular limonite, and the pores in sample S2 are smaller than that of sample S1 (Figure 3c). Sample S3 has fine and uniform grain, loose in texture and there is no adherence of granular limonite (Figure 3d).

Figure 3. Digital photos of deposit samples collected from different locations in rotary kiln: (**a**) S1 with adhered limonite granules on surface; (**b**) characteristics of the longitudinal section of S1; (**c**) characteristics of the longitudinal section of S2; (**d**) appearance characteristics of S3.

3.3. Chemical Composition of Deposits

The chemical compositions of deposit samples S1, S2 and S3 are listed in Table 4. Two results are listed based on the highest and lowest FeO levels of samples S1 and S2, respectively. Compared with the chemical compositions of iron ore (Table 1), it can be seen that the total Fe (TFe) grade of all the deposits increased to about 40% from 35.85% of raw materials, and the main composition of the iron oxides changes from Fe_2O_3 to FeO (wüstite). Analysis shows that the ignition loss of the limonite ore was 15.8% (Table 1); apparently, the TFe grade of the deposits increased as a result of the ignition loss of the limonite. Combined with the chemical compositions of iron ore and coal ash (Tables 1 and 3), the increasing of SiO_2, Al_2O_3, CaO, Na_2O and K_2O contents can be attributed to the addition of coal ash. Previous researches [19–22] indicated that the iron oxides phase of the hematite pellet production deposits was mainly Fe_2O_3 and trace FeO originated from the reduction of hematite by residual carbon particles. The difference is that the iron oxide phase of limonite MRR production deposit was mainly FeO and contains less Fe_2O_3. The reason for the difference might be that the production of hematite pellets are carried out in oxidizing atmosphere; on the contrary, the production of limonite MRR needs to maintain a reduced atmosphere in the rotary kiln by mixed 4–8% coal in limonite. Therefore, the over-reduction reaction occurs to form FeO when limonite passes through high temperature zone in the limonite MRR production. As can be calculated from Table 1, the alkalinity of the iron ore is only about 0.2; thus, the silicate phase can easily form, which may benefit the increase of FeO (wüstite) content. The presence of FeO played a very important role in the formation of the deposits, as it can chemically react with silica and other oxides to form high viscosity eutectic liquid phase [26], reduce the melting point of a system significantly [27] and promotes the generation of liquid phase during the deposit formation.

Table 4. Chemical compositions of deposit samples (wt%).

Deposit Sample	Comp. FeO	TFe	FeO	Fe_2O_3	SiO_2	Al_2O_3	CaO	MgO	MnO	K_2O	Na_2O	S
S1	max	42.51	44.89	8.66	29.57	6.52	2.51	2.69	2.28	2.05	0.15	0.31
	min	39.78	40.16	9.93	30.32	7.64	2.82	2.02	2.09	1.93	0.22	0.27
S2	max	40.67	41.35	9.54	30.18	7.29	2.91	2.25	1.81	2.08	0.21	0.27
	min	38.80	38.32	11.81	31.02	8.00	2.70	2.02	2.05	2.20	0.24	0.13
S3	-	34.53	37.22	6.82	33.42	9.28	4.41	2.40	1.77	2.20	0.32	0.18

The external deposit (S1) possesses more FeO and less Fe_2O_3 than the internal deposit (S2) (Table 4), while other components for both samples are almost the same. This phenomenon might be a result of varied reduction circumstances. For the internal deposit, as it is isolated from the reducing atmosphere, the reduction reaction of iron oxides can only be induced by residual carbon trapped in the deposit. For the external deposit, as it is closer to the flame, the higher temperature causes further over-reduction of iron oxides.

Compared with the deposit on kiln wall (S1, S2), sample S3 contains more SiO_2, Al_2O_3 and CaO and less TFe and FeO. Generally, there are a lot of solid particles contained in the flue gas, mainly composed of coal ash and fine ore particles. There is more coal ash in S3 than in S1 and S2, as the coal ash has lower density than the disturbed limonite ore powder, and the content of coal ash in the flue gas is higher. However, based on the TFe content in Tables 3 and 4, the disturbed limonite ore powder in flue gas also was an important component of sample S3. These particles in the flue gas inertial impact and compress bond to the surface on the top of air duct, chemical reaction occurs in the reducing atmosphere and local high temperature [28,29], and a distinct layered deposition structure was formed with the rotation of the kiln. It can be inferred that the superposition of inertial deposition and chemical reaction was the main reason for the deposition of coal ash and limonite ore powder on the air duct. In addition, as the zone where the air duct located is far away from the flame and has lower temperature, there occurs no visible melting in sample S3. Compared with sample S1 and S2, sample S3 has a more loose structure. It can easily be crushed by manpower, and, when it is large enough, it falls off naturally, which means that it has no remarkable effect on MRR production.

3.4. Microstructural Characterization and Phase Evaluation

The microstructures of the deposit samples at different locations were obtained using a Leica DM4500P polarizing microscope, and the results are shown in Figure 4. As shown in Figure 4, there is a large number of round or elliptical pores inside sample S1 and S2 (Figure 4a,b), while there has a few pores in sample S3 (Figure 4c). The pores in sample S1 (Figure 4b) are not only obvious larger than sample S2 (Figure 4a), but also have many large thin-walled pores. Moreover, there are oriented linear pores only occurs in sample S1 (Figure 4b). The reason for the generation of pores is that limonite ore structure become loose and porous after losing crystal water, then the main component of the reducing gas, carbon monoxide, can be deepened into the limonite ore particles and react with Fe_2O_3 to form CO_2, and those CO_2 bubbles become pores in the solidification of deposits. Generally, the combustion efficiency of the pulverized coal is approximately 80–90% [30]. Therefore, the utilization efficiency of coal in present study may be lower because the coal was mixed in the limonite mainly for limonite reduction under lower temperature. When the coal particles trapped inside the deposits were completely burned or reacted with adjacent limonite particles, the remaining vacancies became pores. In addition, the bubbles in the viscosity liquid phase were elongated when the liquid phase flow and the elongated bubbles became oriented pores if the liquid phase solidifies rapidly (Figure 4b). The rapid solidification of the deposits is possible in limonite MRR production. The production must be stopped to remove the deposits if the deposit ring reached a height that affects the normal operation of the rotary kiln, so the external deposit would undergo a rapid cooling process, resulting in rapid solidification of the liquid phase. Sample S3 (Figure 4c) has loose structure and a few pores, indicating that there was less liquid phase formation therein.

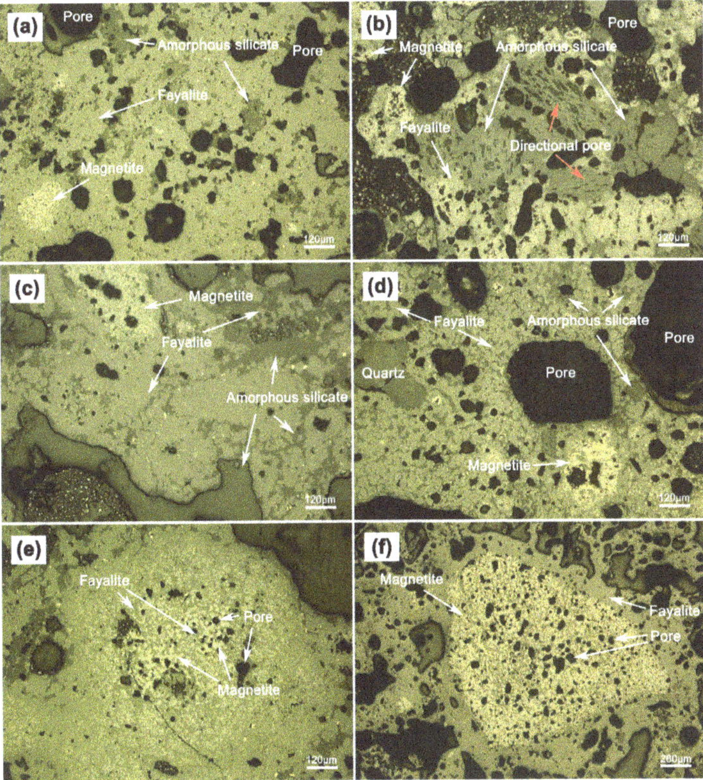

Figure 4. Polarized microscopic images of (**a**) internal deposit (S2); (**b**) external deposit (S1); (**c**) air duct deposit (S3); (**d**) deposit containing independent quartz; (**e**) infiltrated magnetite phase; (**f**) granular magnetite phase.

Figure 4 shows that the phase composition of deposit samples is basically the same. Besides some independent quartz and limonite granules, the deposits consist mainly of fayalite, glass phase (amorphous silicate) and a small amount of magnetite. As the embedded substrate of magnetite, the irregular-shaped fayalite is the most important phase of the deposits at different locations. As shown in Figure 4, it is clear that the fayalite is the main bonding phase, which connects the dispersed magnetite and glass phase. The glass phase is mainly distributed around the pore or disseminates in fayalite, and it has little intersection with magnetite. Magnetite usually embeds in fayalite with a locally enriched in disseminated form (Figure 4e), some are denser round shape or elliptical agglomerates (Figure 4f). The size of magnetite agglomerates varies greatly, generally between 0.02 to 2 mm. The large magnetite agglomerates have obvious boundary, and pores were left by reduction reaction inside. It is clear that these large magnetite agglomerates were mainly derived from the reduced limonite ore (Figure 4f).

The disseminated magnetite agglomerates consist of a large number of spotted worm-like Fe_3O_4 crystallites, indicating that they were crystallized from liquid phase [31].

The structure of mineral phase of deposit samples were determined by XRD analysis. As illustrated in Figure 5, the crystalline phase of S1, S2 and S3 were fayalite, magnetite and quartz. The peak intensity of S1 is lower than that of S2, indicating it contains more amorphous phase in external deposit. Polarization microscopic images (Figure 4a,b) show the same result. Sample S3 has the same phase composition with S1 and S2, but there was more silica in S3, which is consistent with the results of chemical analysis (Table 4).

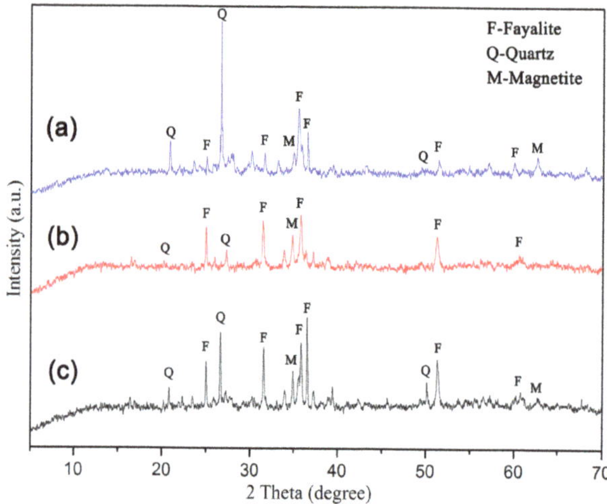

Figure 5. X-ray diffraction (XRD) patterns of: (**a**) air duct deposit (S3); (**b**) external deposit on kiln wall (S1); (**c**) internal deposit on kiln wall (S2).

According to the above analysis, the process of cooling and crystallization of deposits can be inferred. The internal temperature of the deposit ring decreased as the height of the ring increased, and the liquid phase begins to condense and crystallize. According to their melting point, the phase with the highest melting point in the liquid phase precipitates first, while the other phases precipitated sequentially [31]. Possessing the highest melting point (1597 °C) in the system, magnetite precipitated first in the form of metal minerals phase. The iron silicate phase in the deposits then crystallized as fayalite and encapsulated the precipitated magnetite. The phases with low melting point and poor crystallization ability, and the mixture phases which had no time for crystallization in rapid cooling were wrapped around the fayalite and solidified in form of glass phase. Since the melting point of quartz is 1710 °C, it is not easy for the free quartz particles from the raw materials and coal ash to melt amid the MRR process; however, they are prone to be angular erosive. The eroded quartz particles were embedded randomly in the deposit in the form of monomers (Figure 4d).

Due to the serious negative impact on MRR production of deposits on kiln wall, the phase evaluation and elemental distribution of sample S1 and S2 were determined by SEM-EDS analysis. As shown in Figures 6 and 7, the phase composition of the kiln wall deposit is mainly fayalite (point B) combined with magnetite (point A) and glass phase (amorphous silicate, point C), the same on more microscopic scale comparing with the results of polarizing microscopy. As we can see from Figures 6 and 7, the metal phase is almost entirely Fe_3O_4 with few impurities. The content of fayalite is basically the same between different deposits. Compared with other two phases, the glass phase composition is complicated. The content of calcium and silicon therein increased significantly, while the iron content decreased, and small amounts of other elements, including magnesium, aluminum and manganese, are also found in the glass phase.

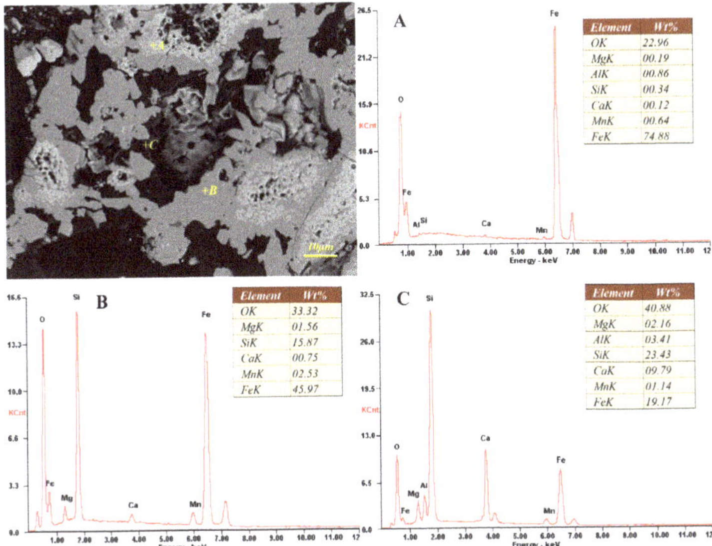

Figure 6. Scanning electron microscopy (SEM)-energy dispersive spectra (EDS) analysis of external deposit sample (S1).

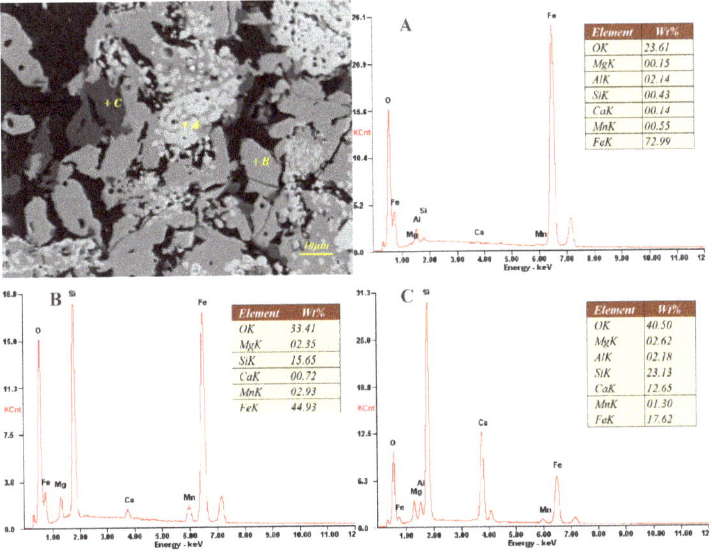

Figure 7. SEM-EDS analysis of internal deposit sample (S2).

As shown in Figure 8, the EDS elemental mapping testified the same elements distribution. It can be seen from Figure 8 that iron is mainly concentrated in magnetite, and secondly in fayalite. Silicon is mainly concentrated in glass phase, and secondary in fayalite. Aluminum and calcium are mainly concentrated in glass phase, while the content of calcium is significantly higher than other phases. Magnesium and manganese exist in a small amount in fayalite and glass phase. For different local deposits, the distribution characteristics of elements are consistent, which suggests that the same rules are followed during solidification of deposits. In addition, Figure 8 shows that there is a small amount of residual carbon left inside the deposits, indicates that it may maintain a reducing condition inside the deposit during its formation, which benefits the formation of FeO.

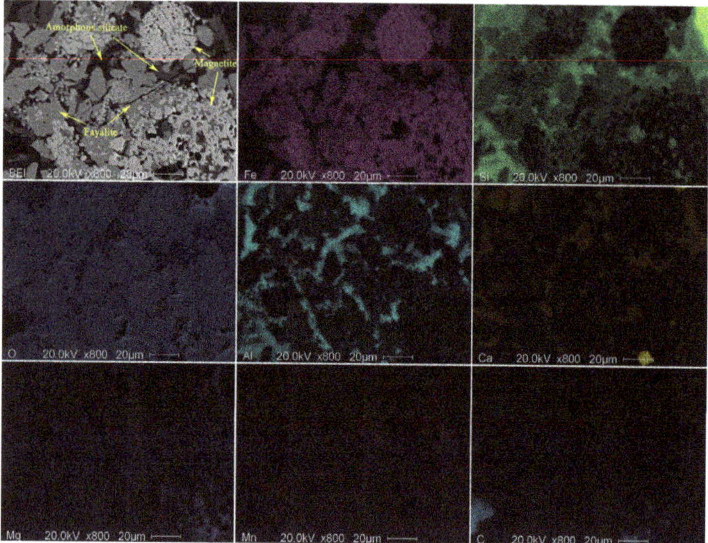

Figure 8. EDS elemental mapping of deposit sample.

3.5. Mechanism for Deposit Formation

3.5.1. Generation of Wüstite (FeO)

Generally, the reduction reactions of coal are regarded as the combination of direct and indirect reduction reactions [32]. Therefore, the reduction reaction of limonite MRR can be regarded as a combination of direct carbon reduction and carbon monoxide reduction. The amount of coal blended in the iron ore of sintering process is directly related to the FeO content in sintered ore. The FeO content in sintered ore can be greater than 30% when the proportion of coal reaches 7% of the iron ore [33]. The amount of coal mixed in the limonite MRR process is 4–8%, which is equivalent to the sintering process. Limonite was easily over-reduced to form FeO by the residual carbon in deposits when the local temperature is too high. It can be seen that most of iron oxides in deposits existed in the form of FeO (see Table 4).

Studied showed formation of FeO begins at about 900 °C from hematite under reduction condition [34]. Even if the FeO content is only 5%, the liquefaction temperature of the low melting point eutectic liquid phase can be reduced to about 920 °C, and the liquid phase content increases with the increase of FeO content [35]. FeO played a very important role for producing low melting point eutectic liquid phase.

3.5.2. Solid Phase Reaction Process of Al_2O_3-SiO_2-CaO-FeO System

The solid-state reaction occurs at the interface of the two particles and produces a corresponding eutectic liquid phase. The temperature at which the solid phase reaction begins is called the Tamman temperature [36], which is much lower than the melting point of the reactants and the system eutectic temperature. The solid state reaction can occur between limonite particles, coal ash particles, or between limonite and coal ash particles in limonite MRR. The particles below 2 mm of the limonite account for about 5–10% in Xinjiang MRR production, which are in a high activation state because of their high lattice defect and high surface free energy. Moreover, decrystallization water and the transformation of the crystal lattice occurred during the reduction of limonite, which leads to partial particle pulverization and strengthens the occurrence of solid-state reaction. In addition, since the alkalinity of the limonite ore is low, the product of solid-state reaction under reducing atmosphere was mainly iron silicate. The results of phase analysis show that fayalite is the most important phase composition and the most important binder phase of the deposits. Alkali metal oxides and salts are known to act corrosively on refractory brick [37]; moreover, the liquid phase may react and bond with the refractory brick [19,38]. These may be two important reasons for the deposits formed and adhered on kiln wall. The iron silicate phase is mainly composed of fayalite, which can be derived from the direct reaction of FeO and SiO_2, or by the reaction of Fe_3O_4 with SiO_2 and CO.

FeO can react with SiO_2 to form low melting point eutectic liquid phase contains fayalite, wüstite and silicate melts [39]. Fe_3O_4, FeO, SiO_2 and solid particles are continuously infiltrated and eroded by the eutectic liquid phase, lead to the continuous expansion of the liquid phase region. The phase diagrams of Al_2O_3-SiO_2-CaO-FeO system with different FeO content were plotted by FactSage® 7.2 software to show the effect of the FeO content on the change of liquid phase region and liquefaction temperature. Figure 9a,b displayed the phase diagrams of the content of FeO were 20% and 40%, respectively. As shown in Figure 9, the gap between the adjacent isotherm lines is 50 °C. Figure 9a shows that the liquefaction temperature is less than 1100 °C when the FeO content is 20%, and it can be seen from Figure 9b that the liquid phase areas significantly expanded and the liquefaction temperature dropped below 1050 °C when the FeO content increased to 40%. Obviously, the amount of low melting point eutectic liquid phase increased while the liquefaction temperature decreases significantly as the FeO content increased.

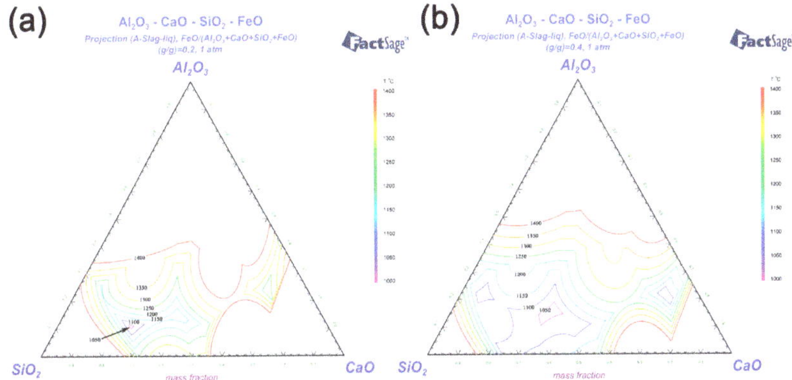

Figure 9. Effects on liquid areas of Al_2O_3-SiO_2-CaO-FeO phase diagrams by FeO of: (**a**) FeO/(Al_2O_3-SiO_2-CaO-FeO) (g/g) = 0.2; (**b**) FeO/(Al_2O_3-SiO_2-CaO-FeO) (g/g) = 0.4.

3.5.3. Analysis of the Deposit Formation Mechanism

The fine particles with larger interfacial energy are more likely to coalesce together, because the presence of low melting point eutectic liquid phase which accelerates the diffusion of the fine-grained particle lattice atoms against the bond force at the mutual contact points or contact surfaces. The aggregated particles wrapped by the eutectic liquid phase exhibit a strong diffusion displacement effect under high temperature, prompting more particles to gather and grow up. Moreover, the eutectic liquid phase will continuously keep melting the high-melting solid particles and expanding the range of liquid phase. The eutectic liquid phase mixed with solid particles adheres to the refractory bricks first [19], and adheres layer by layer with the rotation of the rotary kiln to form layer-structured deposit ring on the kiln wall.

The formation process of air duct deposit was that coal ash and limonite ore particles adhere to the duct head by inertial deposition first, and then bonded by a small amount of low melting point eutectic liquid phase. Therefore, the chemical composition ratio of air duct deposit was slightly different from the kiln wall deposits, and has the same phase composition to the kiln wall deposits because it has the same solid-state reaction and crystallization mechanism. Coal ash contains a large amount of SiO_2 and Al_2O_3, both which can further reduce the alkalinity of raw material, then, liquid phase forms more easily and the formation rate of deposit increases.

The residual carbon retains the reducing atmosphere in deposit ring and promotes the formation of FeO. Compared with other compositions, there are more FeO and SiO_2 in the deposits. These two compositions are the main factors influencing the generating of eutectic liquid phase as they can significantly lower liquefaction point of the deposit. The schematic diagrams of the formation process of kiln wall deposits and the impact deposition mechanism of air duct deposit were shown in Figures 10 and 11, respectively.

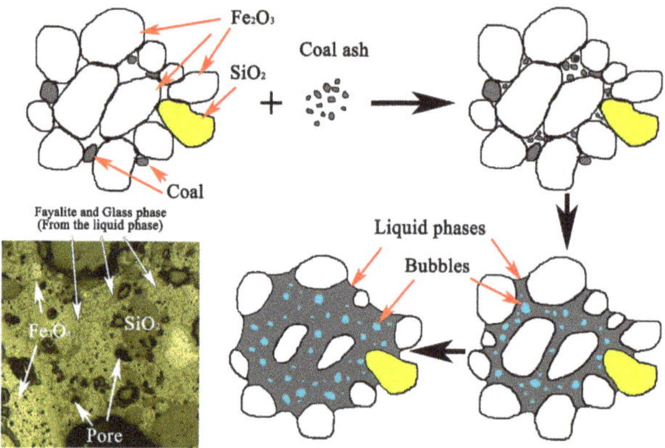

Figure 10. Schematic diagram of kiln wall deposit formation.

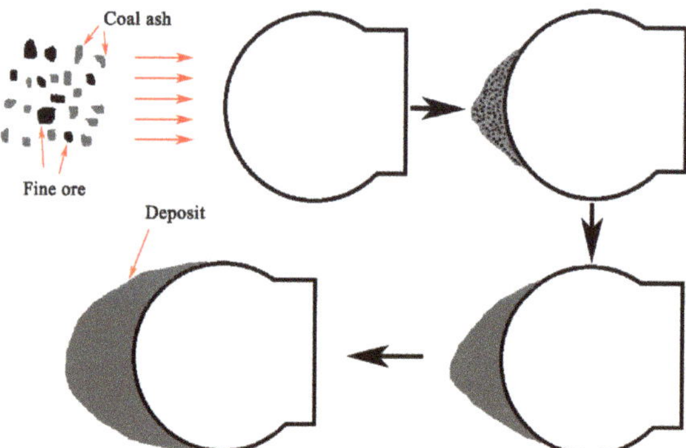

Figure 11. Impact deposition mechanism of air duct deposit.

Influenced by limonite and coal composition, reaction atmosphere and coal ash dosage, the large amount of liquid phase formed amid the MRR production and the formation of deposit is unavoidable. However, the formation rate of the deposit can be significantly slowed down by restraining the formation of low melting point eutectic liquid phase and the over-reduction of limonite, and by reducing the content of coal ash. In the commercial practice in Xinjiang, by reducing the amount of fine-grained limonite and the coal mixed in limonite, or using low-ash coal, the deposit formation rate was reduced to 1/3 to 1/2 of the original value. By moving gas burner timely, the deposits on the kiln wall can periodically be cracked and collapsed, and the non-stop operation time of the rotary kiln MRR system can be significantly prolonged up to 2 months or longer.

4. Conclusions

In this study, the deposits in gas-fired rotary kiln for limonite magnetic-reduction roasting were characterized and the formation mechanism was investigated. The key factors of the deposit formation for the limonite MRR production were clarified.

The deposit on the kiln wall is mainly formed by the solid phase reaction of the fine limonite and coal ash, which generates low melting point eutectic liquid phase adhered granular limonite and contributes to the grown of the deposit. The air duct deposit is mainly caused by the impact deposition of fine particles in flue gas thereon, which results in solid state reaction and liquid phase formation at high temperature, and bonded under the action of the viscous liquid phase.

FeO and SiO_2 are the main factor for generating low melting point eutectic liquid phase. The liquid phase infiltrates and erodes the surrounding solid materials, which results in the continuous expansion of the liquid phase range and causes the solid particles to grow close together. This is the main factor for the rapid growth of the deposit.

The main phase of deposits collected at different locations is mainly fayalite, glass phase, magnetite and a small amount of quartz and limonite. Fayalite, the most important binder phase, accounts for the formation and adhesion of deposits.

According to the deposit formation mechanism of the limonite MRR production, inhibiting the formation of eutectic liquid phase may benefit the slowing down of the development of deposits, and the possible approaches are reducing fine-grained limonite into the kiln, matching the coal blending amount and reducing coal ash into the kiln by using low-ash coal.

Author Contributions: Conceptualization, X.F.; methodology, Y.S.; investigation, X.F. and Z.C.; resources, L.H.; writing-original draft, X.F.; writing-review & editing, X.F. and X.X.; project administration, Z.C.; interpretation of data, L.H.; approved the final version, X.F.

Funding: This research was funded by the 12th five-year National Key Technology Research and Development Program of the Science and Technology of China (No. 2015BAB03B06).

Acknowledgments: The authors wish to thank the colleagues of the Changsha Research Institute of Mining and Metallurgy Co. Ltd for their administrative support and helpful assistance with experiment.

Conflicts of Interest: The authors declare no conflict of interest.

References

1. Wu, F.; Cao, Z.; Wang, S.; Zhong, H. Novel and green metallurgical technique of comprehensive utilization of refractory limonite ores. *J. Clean. Prod.* **2018**, *171*, 831–843. [CrossRef]
2. Liu, G.-S.; Strezovb, V.; Lucas, J.A.; Wibberley, L.J. Thermal investigations of direct iron ore reduction with coal. *Thermochim. Acta* **2004**, *410*, 133–140. [CrossRef]
3. Luo, L.; Zhang, J.; Yu, Y. Recovering limonite from Australia iron ores by flocculation-high intensity magnetic separation. *J. Cent. South Univ. Technol.* **2005**, *12*, 682–687. [CrossRef]
4. Chen, W.; Liu, X.; Peng, Z.; Wang, Q. Recovery of Huangmei limonite by flash magnetic roasting technique. In Proceedings of the 3rd International Symposium on High-Temperature Metallurgical Processing, Hoboken, NJ, USA, 17 March 2012; pp. 49–58.
5. Rath, S.S.; Dhawan, N.; Rao, D.S.; Das, B.; Mishra, B.K. Beneficiation studies of a difficult to treat iron ore using conventional and microwave roasting. *Powder Technol.* **2016**, *301*, 1016–1024. [CrossRef]
6. Nasr, M.; Youssef, M.A. Optimization of magnetizing reduction and magnetic separation of iron ores by experimental design. *ISIJ Int.* **1996**, *36*, 631–639. [CrossRef]
7. Youssef, M.A.; Morsi, M.B. Reduction roast and magnetic separation of oxidized iron ores for the production of blast furnace feed. *Can. Metall. Q.* **2013**, *37*, 419–428. [CrossRef]

8. Li, C.; Sun, H.; Bai, J.; Li, L. Innovative methodology for comprehensive utilization of iron ore tailings: Part 1. The recovery of iron from iron ore tailings using magnetic separation after magnetizing roasting. *J. Hazard. Mater.* **2010**, *174*, 71–77. [CrossRef]

9. Rath, S.S.; Sahoo, H.; Dhawan, N.; Rao, D.S.; Das, B.; Mishra, B.K. Optimal recovery of iron values from a low grade iron ore using reduction roasting and magnetic separation. *Sep. Sci. Technol.* **2014**, *49*, 1927–1936. [CrossRef]

10. Shuai, X.; Zhang, Y.; Xu, H.; Zhou, H.; Qi, Y. Magnetic roasting of limonite of low grade. *J. Iron Steel Res.* **2015**, *27*, 8–12. (In Chinese)

11. Ponomar, V.P.; Dudchenko, N.O.; Brik, A.B. Synthesis of magnetite powder from the mixture consisting of siderite and hematite iron ores. *Miner. Eng.* **2018**, *122*, 277–284. [CrossRef]

12. Zhu, Q.; Li, H. Status quo and development prospect of magnetizing roasting via fluidized bed for low grade iron ore. *CIESC J.* **2014**, *65*, 2437–2442. (In Chinese) [CrossRef]

13. Yan, A.; Chai, T.; Yu, W.; Xu, Z. Multi-objective evaluation-based hybrid intelligent control optimization for shaft furnace roasting process. *Control Eng. Pract.* **2012**, *20*, 857–868. [CrossRef]

14. Wang, Q.; Chen, W.; Yu, Y.; Yan, X.; Liu, X. Test research on the flash magnetization roasting technology for complex and refractory iron ore. *Metal Mine* **2009**, *402*, 73–76. (In Chinese)

15. Li, Y.; Zhu, T. Recovery of low grade haematite via fluidised bed magnetising roasting: Investigation of magnetic properties and liberation characteristics. *Ironmak. Steelmak.* **2013**, *39*, 112–120. [CrossRef]

16. Liu, X.; Yu, Y.; Hong, Z.; Peng, Z.; Li, J.; Zhao, Q. Development and application of packaged technology for flash(fluidization) magnetizing roasting of refractory weakly magnetic iron ore. *Mining Metall. Eng.* **2017**, *37*, 40–45. (In Chinese) [CrossRef]

17. Zhang, X.; Han, Y.; Sun, Y.; Li, Y. Innovative utilization of refractory iron ore via suspension magnetization roasting: A pilot-scale study. *Powder Technol.* **2019**, *352*, 16–24. [CrossRef]

18. Fu, X.; Mao, Y.; Xue, S. Study on deposit formation and prevention control of large industrial magnetized roasting rotary kiln. *Nonferrous Met. (Miner. Process. Section)* **2013**, *s1*, 236–239. (In Chinese)

19. Wang, S.; Guo, Y.; Fan, J.; He, Y.; Jiang, T.; Chen, F.; Zheng, F.; Yang, L. Initial stage of deposit formation process in a coal fired grate-rotary kiln for iron ore pellet production. *Fuel Process. Technol.* **2018**, *175*, 54–63. [CrossRef]

20. Wang, S.; Guo, Y.; Fan, J.; Jiang, T.; Chen, F.; Zheng, F.; Yang, L. Deposits in a coal fired grate-kiln plant for hematite pellet production: Characterization and primary formation mechanisms. *Powder Technol.* **2018**, *333*, 122–137. [CrossRef]

21. Guo, Y.-F.; Wang, S.; He, Y.; Jiang, T.; Chen, F.; Zheng, F.-Q. Deposit formation mechanisms in a pulverized coal fired grate for hematite pellet production. *Fuel Process. Technol.* **2017**, *161*, 33–40. [CrossRef]

22. Zhong, Q.; Yang, Y.; Jiang, T.; Li, Q.; Xu, B. Effect of coal ash on ring behavior of iron-ore pellet powder in kiln. *Powder Technol.* **2018**, *323*, 195–202. [CrossRef]

23. Jiang, T.; He, G.; Gan, M.; Li, G.; Fan, X.; Yuan, L. Forming mechanism of rings in rotary-Kiln for forming mechanism of rings in rotary-kiln for oxidized pellet. *J. Iron Steel Res. Int.* **2009**, *16*, 292–297.

24. Stjernberg, J.; Jonsson, C.Y.C.; Wiinikka, H.; Lindblom, B.; Boström, D.; Öhman, M. Deposit formation in a grate–kiln plant for iron-ore pellet production. Part 2: Characterization of deposits. *Energy Fuels* **2013**, *27*, 6171–6184. [CrossRef]

25. Zheng, G. Chemical analysis method standards and proficiency testing of laboratory for iron ore. *Metall. Anal.* **2015**, *35*, 37–44. (In Chinese)

26. Wu, X.; Ji, H.; Dai, B.; Zhang, L. Xinjiang lignite ash slagging and flowability under the weak reducing environment at 1300 °C—A new method to quantify slag flow velocity and its correlation with slag properties. *Fuel Process. Technol.* **2018**, *171*, 173–182. [CrossRef]

27. Zhang, Z.; Wu, X.; Zhou, T.; Chen, Y.; Hou, N.; Piao, G.; Kobayashi, N.; Itaya, Y.; Mori, S. The effect of iron-bearing mineral melting behavior on ash deposition during coal combustion. *Proc. Combust. Inst.* **2011**, *33*, 2853–2861. [CrossRef]

28. Baxter, L.L.; DeSollart, R.W. A mechanistic description of ash deposition during pulverized coal combustion predictions compared with observations. *Fuel* **1993**, *72*, 1411–1418. [CrossRef]

29. Baxter, L.L. Ash deposition during biomass and coal combustion a mechanistic approach. *Biomass Bioenerg.* **1993**, *4*, 85–102. [CrossRef]

30. Ishii, K. (Ed.) *Advanced Pulverized Coal Injection Technology and Blast Furnace Operation*; Elsevier: Amsterdam, The Netherlands, 2000.
31. Chen, Y.; Chen, R. *Microstructure of Sinter and Pellet*; Central South University Press: Changsha, China, 2011. (In Chinese)
32. Ma, B.; Wang, C.; Yang, W.; Yin, F.; Chen, Y. Screening and reduction roasting of limonitic laterite and ammonia-carbonate leaching of nickel–cobalt to produce a high-grade iron concentrate. *Miner. Eng.* **2013**, *50*, 106–113. [CrossRef]
33. Zhang, H. *Theory and Technology of Sintered Pellets*; Chemical Industry Press: Beijing, China, 2015. (In Chinese)
34. Zeng, T.; Helble, J.J.; Bool, L.E.; Sarofim, A.F. Iron transformations during combustion of Pittsburgh No. 8 coal. *Fuel* **2009**, *88*, 566–572. [CrossRef]
35. Wang, S.; Guo, Y.; Chen, F.; He, Y.; Jiang, T.; Zheng, F. Combustion reaction of pulverized coal on the deposit formation in the kiln for iron ore pellet production. *Energy Fuels* **2016**, *30*, 6123–6131. [CrossRef]
36. Merkle, R.; Maier, J. On the tammann-rule. *Z. Anorg. Allg. Chem.* **2005**, *631*, 1163–1166. [CrossRef]
37. Dahotre, N.B.; Kadolkar, P.; Shah, S. Refractory ceramic coatings: Processes, systems and wettability/adhesion. *Surf. Interface Anal.* **2001**, *31*, 659–672. [CrossRef]
38. Stjernberg, J.; Ion, J.C.; Antti, M.-L.; Nordin, L.-O.; Lindblom, B.; Odén, M. Extended studies of degradation mechanisms in the refractory lining of a rotary kiln for iron ore pellet production. *J. Eur. Ceram. Soc.* **2012**, *32*, 1519–1528. [CrossRef]
39. Huffman, G.P.; Huggins, F.E.; Dunmyre, G.R. Investigation of the high-temperature behaviour of coal ash in reducing and oxidizing atmospheres. *Fuel* **1981**, *60*, 585–597. [CrossRef]

Article

Modification Mechanism of Spinel Inclusions in Medium Manganese Steel with Rare Earth Treatment

Zhe Yu [1,2] and Chengjun Liu [1,2,*]

[1] Key Laboratory for Ecological Metallurgy of Multimetallic Ores (Ministry of Education), Shenyang 110819, China
[2] School of Metallurgy, Northeastern University, Shenyang 110819, China
* Correspondence: liucj@smm.neu.edu.cn; Tel.: +86-138-9885-0333

Received: 7 July 2019; Accepted: 19 July 2019; Published: 21 July 2019

Abstract: In aluminum deoxidized medium manganese steel, spinel inclusions are easily to form during refining, and such inclusions will deteriorate the toughness of the medium manganese steel. Rare earth inclusions have a smaller hardness, and their thermal expansion coefficients are similar to that of steel. They can avoid large stress concentrations around inclusions during the heat treatment of steel, which is beneficial for improving the toughness of steel. Therefore, rare earth Ce is usually used to modify spinel inclusions in steel. In order to clarify the modification mechanism of spinel inclusions in medium manganese steel with Ce treatment, high-temperature simulation experiments were carried out. Samples were taken step by step during the experimental steel smelting process, and the inclusions in the samples were analyzed by SEM-EDS. Finally, the experimental results were discussed and analyzed in combination with thermodynamic calculations. The results show that after Ce treatment, the amount of inclusions decrease, the inclusion size is basically less than 5 μm, and the spinel inclusions are transformed into rare earth inclusions. After Ce addition, Mn and Mg in the spinel inclusions are first replaced by Ce, and the spinel structure is destroyed to form $CeAlO_3$. When the O content in the steel is low, S in the steel will replace the O in the inclusion, and $CeAlO_3$ and spinel inclusions will be transformed into Ce_2O_2S. By measuring the total oxygen content of the steel, the total Ce content required for complete modification of spinel inclusions can be obtained. Finally, the critical conditions for the formation and transformation of inclusions in the Fe-Mn-Al-Mg-Ce-O-S system at 1873K were obtained according to thermodynamic calculations.

Keywords: medium manganese steel; spinel inclusions; Ce treatment; modification mechanism

---

1. Introduction

Medium manganese steel is a typical representative of third generation automotive steel with high strength and high plasticity. It can be applied to many body structure parts including automobile A-pillars, B-pillars, anticollision beams, and sill reinforcement. Industrial experimental studies have shown that $(Mn,Mg)O·Al_2O_3$- and $MgO·Al_2O_3$-type spinel inclusions will form during refining of aluminum deoxidized medium manganese steel [1]. Spinel inclusions are generally angular and have high hardness, which is the main cause of stamping and cracking defects in automotive steel [2,3]. Therefore, it is necessary to modify the spinel inclusions to improve the toughness of medium manganese steel. At present, low-melting CaO-MgO-Al_2O_3 inclusions obtained by calcium treatment are generally expected to avoid the damage of hard inclusions [4,5]. However, large-size inclusions cannot be completely modified due to kinetic conditions, which means that the spinel phase core of the complex inclusions will still exist and be exposed after rolling [6].

The rare earth element Ce has a good modification effect on the spinel [7–9], which can modify the hard oxide inclusions into rare earth inclusions with a soft hardness [10]. Huang et al. [7,9] and Wang et al. [8] respectively studied the modification behavior of Ce on inclusions in drill steel and high-speed railway steel, and the transformation mechanism of spinel to Ce_2O_3 and Ce_2O_2S was described. However, there are relatively few reports on rare earth modified spinel inclusions. The modification mechanism of Ce on the spinel inclusions is not comprehensive, and the effect of aluminum content on Ce treatment has not been reported. Therefore, in this paper, the modification behaviors of spinel inclusions after Ce treatment in medium manganese steel with different aluminum contents were studied. Meanwhile, the Ce content required for complete modification of spinel inclusions was discussed in combination with thermodynamic calculations, which are intended to guide the appropriate addition of rare earth in the actual production process.

2. Experimental Procedure

High-temperature simulation experiments were carried out by using a $MoSi_2$ furnace (TianJin Muffle Technology Co. Ltd, Tianjin, China). According to the general composition of medium manganese steel, the target Al content was set to three levels: low aluminum (0.02%), numbered R1; medium aluminum (0.2%), numbered R2; and high aluminum (2.0%), numbered R3. The experimental steps are as follows. Approximately 400 g of pure iron and an MnFe alloy was placed in a corundum crucible and then placed in the constant temperature zone of the furnace. The master irons (pure iron and MnFe alloy) were heated to 1873 K under argon atmosphere with a flow rate of 0.4 L min^{-1}. Subsequently, the oxygen activity in the molten steel was measured by an electrolyte oxygen probe, and it was defined as the initial oxygen activity (No. 1 $a_{[O]}$). Based on the target Al content, a certain amount of Al wire was added, and NiMg alloy was added 10 min later. No. 2 $a_{[O]}$ was measured by electrolyte oxygen probe after another 10 min from the addition of NiMg alloy, and then sample 0 was taken by a quartz tube and quenched in an ice bath. Finally, CeFe alloy was added into the molten steel to modify inclusions. Thereafter, the molten steel was respectively held for 1, 10, and 30 min at 1873 K (1600 °C) and then successively sampled as above (sample 1–sample 3). After taking sample 2, the oxygen activity was measured by an electrolyte oxygen probe, which was defined as the final oxygen activity (No. 3 $a_{[O]}$). The experimental process is provided in Figure 1. Table 1 shows the compositions of raw materials, including pure iron, MnFe alloy, Al wire, NiMg alloy, and CeFe alloy.

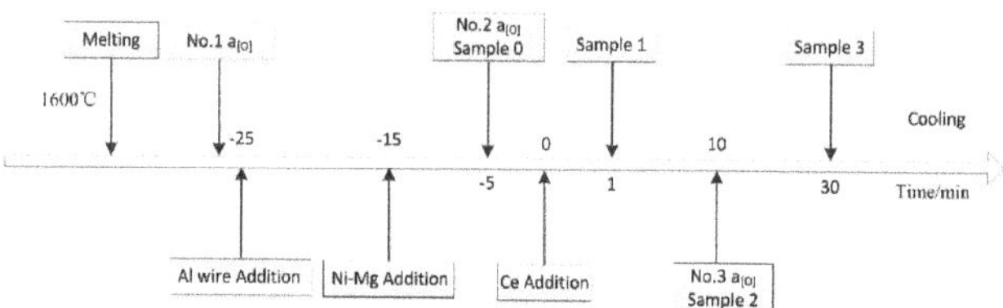

Figure 1. Schematic diagram of the experimental process.

Table 1. Chemical composition of raw materials (mass fraction/%).

Type	Fe	Ni	Mg	Al	C	Si	Mn	Ce	S	P	Others
Pure iron	99.9	0.01	-	0.001	0.0021	0.01	0.03	-	0.003	0.005	0.03
Al wire	-	-	-	99.99	-	-	-	-	-	-	0.01
MnFe alloy	12.57	-	-	-	1.5	1.5	84.2	-	0.03	0.2	-
NiMg alloy	0.93	80.12	17.98	-	0.78	0.19	-	-	-	-	-
CeFe alloy	79	-	-	-	-	-	-	19	-	-	2

To observe inclusions of steel, the steel samples were polished by SiC paper and diamond suspensions. The chemical compositions and morphology of inclusions were analyzed through a scanning electron microscope (SEM, Phenom Prox, Eindhoven, The Netherlands) and an energy dispersive spectroscope (EDS, Phenom Prox, Eindhoven, The Netherlands). The contents of aluminum, manganese, magnesium, and cerium in the steel were determined by using an inductively coupled plasma optical emission spectrometer (ICP-OES, PerkinElmer, Waltham, MA, USA). The content of sulfur was measured by a carbon and sulfur analyzer. The total oxygen content was measured by an oxygen and nitrogen analyzer. The contents of each element in the three steels (numbered R1, R2, and R3) are listed in Table 2.

Table 2. Chemical composition of experimental steels (mass fraction/%).

No.	C	Mn	Al	Mg	Ce	S	$a_{[O]}$			TO
							No. 1	No. 2	No. 3	
R1	0.23	8.95	0.019	0.001	0.014	0.016	0.0149	0.0014	0.0005	0.0045
R2	0.21	9.18	0.21	0.001	0.015	0.016	0.0124	0.0002	0.0002	0.0021
R3	0.25	9.35	2.03	0.002	0.013	0.014	0.0149	0.0001	0.0001	0.0006

3. Results

3.1. Inclusion Transformation

The types of typical inclusions in the experimental steel are shown in Figure 2; there are five types in common. The type 1 inclusions are Al_2O_3, as shown in Figure 2a. Almost no obvious Mn element was detected in this type of inclusion. Both the type 2 and type 3 are spinel inclusions, namely $(Mn,Mg)O \cdot Al_2O_3$ and $MgO \cdot Al_2O_3$ spinel, respectively. The type 4 and type 5 inclusions are rare earth inclusions, and they appear bright white under SEM backscatter conditions. The type 4 inclusions are $CeAlO_3$, which are mostly block-shaped and have a small size of about 2~5 µm. The type 5 inclusions are Ce_2O_2S, which are mostly spherical and have a size of about 3~8 µm.

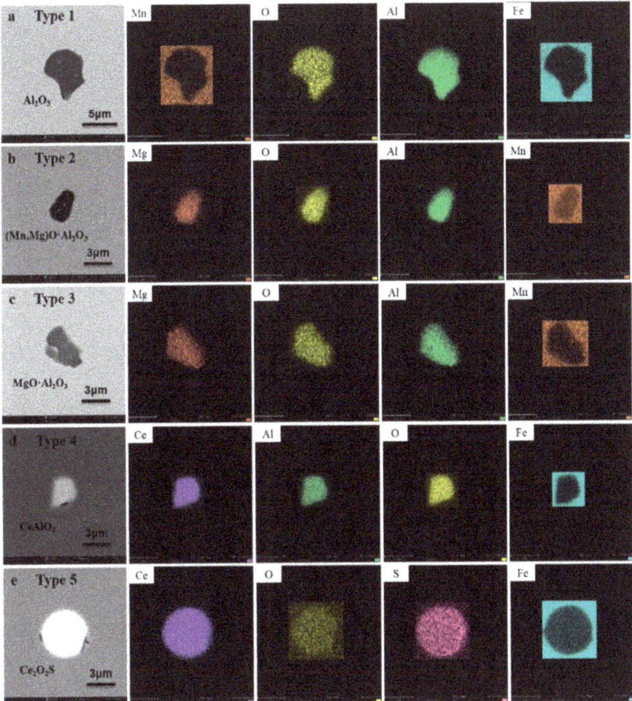

Figure 2. Elemental mapping of typical inclusions in steel: (**a**) Al_2O_3; (**b**) $(Mn,Mg)O \cdot Al_2O_3$; (**c**) $MgO \cdot Al_2O_3$; (**d**) $CeAlO_3$; (**e**) Ce_2O_2S.

After adding Ce, in addition to the above inclusions with uniform composition, there was a small amount of complex inclusions in the steel that underwent modification, as shown in Figure 3. In Figure 3a, the inclusion core is Mg-Al-O, and the outer layer is Ce-Al-O. In Figure 3b, the inclusion core is Mg-Al-O, and the outer layer is Ce-O-S. In Figure 3c, the inclusion is divided into three layers. The core is Mg-Al-O, the middle layer is Ce-Al-O, and the outermost layer is Ce-O-S. In Figure 3d, the inclusion core is Ce-O-S, and the outer layer is Ce-Al-O.

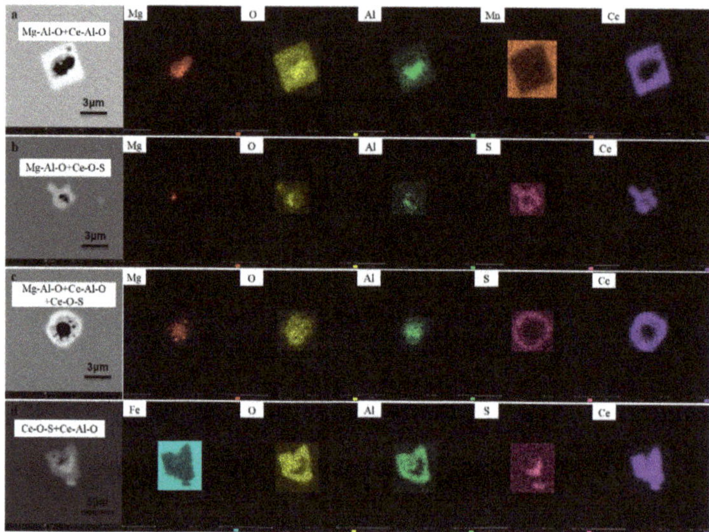

Figure 3. Elemental mapping of typical complex inclusions: (**a**) Mg-Al-O + Ce-Al-O; (**b**) Mg-Al-O + Ce-O-S; (**c**) Mg-Al-O + Ce-Al-O + Ce-O-S; (**d**) Ce-O-S + Ce-Al-O.

The types of uniform inclusions and complex inclusions for each sampling period are listed in Table 3. It can be seen that the inclusions in the three experimental steels were mainly spinel after aluminum deoxidation and magnesium treatment, and there was also a small amount of Al_2O_3 inclusions. The R1 experimental steel mainly contained $(Mn,Mg)O \cdot Al_2O_3$ inclusions. The spinel inclusions in the R2 and R3 experimental steels were mainly $MgO \cdot Al_2O_3$.

Table 3. Types of inclusions at different stages.

Type	Chemical Formula of Inclusions	Samples in R1 Steel				Samples in R2 Steel				Samples in R3 Steel			
		0	1	2	3	0	1	2	3	0	1	2	3
1	Al_2O_3	√				√				√			
2	$(Mn,Mg)O \cdot Al_2O_3$	√√											
3	$MgO \cdot Al_2O_3$	√				√√				√			
4	$CeAlO_3$		√√	√√	√√	√√	√	√					
5	Ce_2O_2S		√√	√		√√	√√	√√			√	√	√
a	Mg-Al-O + Ce-Al-O		√	√		√	√						
b	Mg-Al-O + Ce-O-S		√			√	√				√		
c	Mg-Al-O + Ce-Al-O + Ce-O-S		√			√					√		
d	Ce-O-S + Ce-Al-O			√									

Note: √√ represents the main type of inclusions, and √ represents a small number of inclusions.

After Ce treatment for 1 min, the inclusions in R1 experimental steel were mainly $CeAlO_3$ and Ce_2O_2S, and there was a small amount of complex inclusions of type a (Mg-Al-O + Ce-Al-O), b (Mg-Al-O + Ce-OS), and c (Mg-Al-O + Ce-Al-O + Ce-OS). After Ce treatment for 10 min, the inclusions in steel were mainly $CeAlO_3$, while there was still a small amount of Ce_2O_2S inclusions as well as type a (Mg-Al-O + Ce-Al-O) and d (Ce-O-S + Ce-Al-O) complex inclusions. After Ce treatment for 30 min, the inclusions in the steel were almost all $CeAlO_3$. It indicated that Ce-Al-O and Ce-O-S inclusions would be directly formed in the steel after Ce addition. Meanwhile, spinel inclusions would also be modified by Ce. As time prolonged, spinel inclusions were gradually modified completely, and Ce-O-S would be transformed into Ce-Al-O.

After Ce treatment for 1 min, the inclusions in R2 experimental steel were mainly $CeAlO_3$ and Ce_2O_2S, and there was a small amount of complex inclusions of type a (Mg-Al-O + Ce-Al-O), b (Mg-Al-O + Ce-OS), and c (Mg-Al-O + Ce-Al-O + Ce-OS). After Ce treatment for 10 min, the inclusions in steel were mainly $CeAlO_3$, while there was still a small amount of Ce_2O_2S inclusions as well as type a (Mg-Al-O + Ce-Al-O) and b (Mg-Al-O + Ce-O-S) complex inclusions. After Ce treatment for 30 min, both $CeAlO_3$ and Ce_2O_2S were present in the steel. It indicated that Ce-Al-O and Ce-O-S inclusions would directly form in the steel after Ce addition. Meanwhile, spinel inclusions would also be modified by Ce. As time prolonged, spinel inclusions would gradually be transformed into Ce-Al-O and Ce-O-S.

After Ce treatment for 1 min, the inclusions in R3 experimental steel were mainly Ce_2O_2S, and there was a small amount of complex inclusions of type a (Mg-Al-O + Ce-Al-O) and c (Mg-Al-O + Ce-Al-O + Ce-OS). After Ce treatment for 10 min, the inclusions in steel were mainly Ce_2O_2S. It indicated that Ce-O-S inclusions would directly form in the steel after Ce addition. Meanwhile, spinel inclusions would also be modified by Ce. As time prolonged, spinel inclusions would be gradually transformed into Ce-O-S.

3.2. The Number Density and Size Change of Inclusions

In order to investigate the influence of Ce treatment on the number and size of inclusions in steel, Image pro plus software (Version 6.0, Media Cybernetics, Shanghai, China) was used to count the number density (number of inclusions per square millimeter) and size (equivalent diameter) of inclusions for each sampling period in the three experimental steels. Each sample counted 150 fields of view at 2000 magnification, for a total area of 1.5 mm^2. The number density and size change of inclusions during the experiment are shown in Figure 4.

It can be seen from Figure 4 that the variation law of the size and number density of inclusions in the three experimental steels is basically identical. After Ce treatment for 1 min, the proportion of inclusions with a size 0.5–2.0 μm (see Figure 4a) and the number density of inclusions (see Figure 4b) increased significantly. With the prolongation of treatment time, the number density of inclusions showed a decreasing trend, and the size of inclusions showed a slight increasing trend. However, inclusions with a size less than 5 μm were still dominant after 30 min.

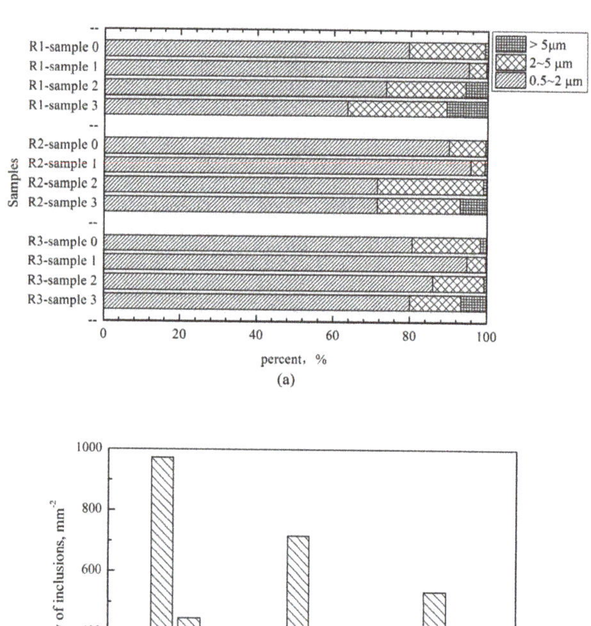

Figure 4. Size distribution (**a**) and number density (**b**) of inclusions in three steels at different stages.

4. Thermodynamics Analysis

After Ce treatment, the oxygen activity in the R1 experimental steel decreased (see Table 2), and the inclusions in the steel were mainly $CeAlO_3$ at 30 min after Ce treatment (see Table 3). The oxygen activity in R2 and R3 steels before Ce treatment had dropped to a very low level due to the high Al content. Therefore, the oxygen activity would not be further decreased after Ce addition (see Table 2). The inclusions present in the steels were mainly Ce_2O_2S at 30 min after Ce treatment, and only a small amount of $CeAlO_3$ inclusions were found in the R2 experimental steel (see Table 3). It indicates that Ce treatment can modify spinel inclusions into rare earth inclusions. The deoxidation reaction between Ce and O will preferentially proceed after the addition of Ce to steel, and Ce will react with S when the O content is at a very low level. The experimental phenomenon is consistent with previous research results [11,12].

To assess the evolution mechanism of inclusions in the steel, an inclusion stability diagram of Fe-10%Mn-0.001%Mg-0.015%S-0.015%Ce-Al-O system at 1873 K was calculated by FactSage™ program (Version 7.1, developed by Thermfact/CRCT (Montreal, Canada) and GTT-Technologies (Achen, Germany)), as shown in Figure 5. The compositions of the three experimental steels are also marked in it. In the calculation, in addition to FactPS, FToxid, and FSstel databases, a privately created database MYDT containing thermodynamic data of Ce_2O_2S [13] was also selected. As can be seen from the figure, the stable inclusion region shows a change of $CeAlO_3 \rightarrow CeAlO_3 + Ce_2O_2S \rightarrow Ce_2O_2S$ as the oxygen content decreases within a fairly large range of Al content in the steel. The composition of R1 experimental steel was located in the CeAlO$_3$ stable region, and the stable inclusions in the R2 experimental steel were CeAlO$_3$ + Ce$_2$O$_2$S, which was consistent with the experimental results (see Table 3). The thermodynamic calculation results of R3 experimental steel showed that there were no inclusions, which was inconsistent with the experimental results. The main reason is that thermodynamic calculations by Factsage have large deviations from the actual situation when the Al content is higher. This phenomenon is also proved by others [14], and a large amount of data indicate that the actual equilibrium oxygen content in steel is lower than the calculated value under high aluminum conditions in steel. Therefore, the activity relationship will be used to express the critical condition of inclusion transformation as follows.

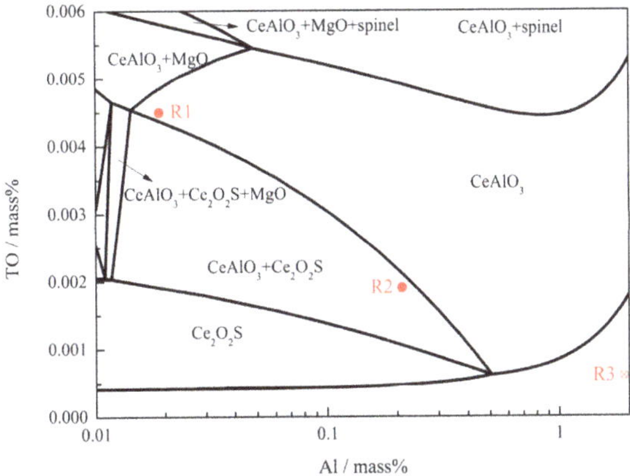

Figure 5. Stability diagram in the Fe-10%Mn-0.001%Mg-0.015%S-0.015%Ce-Al-O system at 1873 K.

According to the experimental results, the chemical reactions involved in the Fe-Mn-Al-Mg-Ce-O-S system and related standard Gibbs free energy [15–20] are shown in Table 4.

Table 4. Possible chemical reactions in steel and its standard Gibbs free energy.

No.	Reaction Equations in Molten Steel	$\Delta G^{\ominus}$ (J/mol)
1	2[Al] + 3[O] = Al$_2$O$_3$(s)	−867,370 + 222.5T [20]
2	[Mg] + [O] = MgO(s)	−89,960 − 82.0T [19]
3	[Mn] + [O] = MnO(s)	−287,900 + 125T [17]
4	[Mn] + 2[Al] + 4[O] = MnO·Al$_2$O$_3$(s)	−1,545,000 + 524T [17]
5	[Mg] + 2[Al] + 4[O] = MgO·Al$_2$O$_3$(s)	−978,182 + 128.93T [18]
6	2[Ce] + 2[O] + [S] = Ce$_2$O$_2$S(s)	−1,353,592.4 + 331.6T [16]
7	[Ce] + [Al] + 3[O] = CeAlO$_3$(s)	−1,366,460 + 364T [15]

The deoxidation equilibrium relationship of the molten steel includes Al-O, Al-Mg-O, Al-Ce-O, and Ce-S-O, and the formation conditions of the corresponding inclusions at 1873 K are as follows.

The formation condition of Al$_2$O$_3$ inclusions is that oxygen activity in steel is greater than that limited by the Al–O equilibrium, as shown in Equation (1)

$$a_O > a_{Al}^{-2/3} \cdot K_{Al_2O_3}^{-1/3} = 6.468 \times 10^{-5} \cdot a_{Al}^{-2/3}, \tag{1}$$

where K represents the equilibrium constant of the corresponding reaction, and a represents the activity of the corresponding component (the same as below).

The formation condition of $MgO \cdot Al_2O_3$ inclusions is that the actual oxygen activity in steel is greater than that limited by the Mg-Al-O equilibrium, as shown in Equation (2)

$$a_O > a_{Al}^{-1/2} \cdot a_{Mg}^{-1/4} \cdot K_{MgO \cdot Al_2O_3}^{-1/4} = 7.303 \times 10^{-6} \cdot a_{Al}^{-1/2} \cdot a_{Mg}^{-1/4}. \tag{2}$$

The formation conditions of $CeAlO_3$ inclusions is that the actual oxygen activity in steel is greater than that limited by Ce-Al-O equilibrium, as shown in Equation (3)

$$a_O > a_{Al}^{-1/3} \cdot a_{Ce}^{-1/3} \cdot K_{CeAlO_3}^{-1/3} = 4.31 \times 10^{-7} \cdot a_{Al}^{-1/3} \cdot a_{Ce}^{-1/3}. \tag{3}$$

The formation conditions of Ce_2O_2S inclusions is that the actual oxygen activity in steel is greater than that limited by Ce-O-S equilibrium, as shown in Equation (4)

$$a_O > a_{Ce}^{-1} \cdot a_S^{-1/2} \cdot K_{Ce_2O_2S}^{-1/2} = 6.1 \times 10^{-11} \cdot a_{Ce}^{-1} \cdot a_S^{-1/2}. \tag{4}$$

The transformation reaction between the inclusions can be obtained by associating the reactions in Table 2, as follows.

Conditions for Al_2O_3 transformation into $MgO \cdot Al_2O_3$:

$$4Al_2O_3(s) + 3[Mg] = 2[Al] + 3MgO \cdot Al_2O_3(s) \quad \Delta G^\theta = 534,934 - 503.21T. \tag{5}$$

Let $\Delta G < 0$, at 1873 K:

$$a_{Mg} > 1.625 \times 10^{-4} \cdot a_{Al}^{2/3}. \tag{6}$$

Conditions for $MgO \cdot Al_2O_3$ transformation into $CeAlO_3$:

$$4[Ce] + 3MgO \cdot Al_2O_3(s) = 4CeAlO_3(s) + 3[Mg] + 2[Al] \quad \Delta G^\theta = -2,531,294 + 1069.21T. \tag{7}$$

Let $\Delta G < 0$, at 1873 K:

$$a_{Ce} > 2.061 \times 10^{-4} \cdot a_{Mg}^{3/4} \cdot a_{Al}^{1/2}. \tag{8}$$

Conditions for $CeAlO_3$ transformation into Ce_2O_2S:

$$[Ce] + [S] + CeAlO_3(s) = Ce_2O_2S(s) + [Al] + [O] \quad \Delta G^\theta = 12,867.6 - 32.4T. \tag{9}$$

Let $\Delta G < 0$, at 1873 K:

$$a_{Ce} > 0.0464 \cdot a_O \cdot a_{Al} \cdot a_S^{-1}. \tag{10}$$

Conditions for $MgO \cdot Al_2O_3$ transformation into Ce_2O_2S:

$$8[Ce] + 4[S] + 3MgO \cdot Al_2O_3(s) = 4Ce_2O_2S(s) + 3[Mg] + 6[Al] + 4[O]$$
$$\Delta G^\theta = -2,479,823.6 + 939.61T. \tag{11}$$

Let $\Delta G < 0$, at 1873 K:

$$a_{Ce} > 3.09 \times 10^{-3} \cdot a_{Mg}^{3/8} \cdot a_{Al}^{3/4} \cdot a_O^{1/2} \cdot a_S^{-1/2}. \tag{12}$$

According to the above analysis, the inclusion formation conditions can be obtained as follows.

(1) The formation condition of $MgO \cdot Al_2O_3$ in steel at 1873 K after magnesium treatment needs to be considered from two different formation mechanisms, as above mentioned. One is that Al_2O_3 can transform into $MgO \cdot Al_2O_3$ (as shown in Equation (6)), which will provide a limit on magnesium activity. The other one is that $MgO \cdot Al_2O_3$ can directly form from steel (as shown in Equation (2)), which requires the oxygen activity to be higher than the equilibrium oxygen activity limited by Mg-Al-O, as follows.

$$a_{Mg} > 1.625 \times 10^{-4} \cdot a_{Al}^{2/3}, \quad a_O > 7.303 \times 10^{-6} \cdot a_{Al}^{-1/2} \cdot a_{Mg}^{-1/4}. \tag{13}$$

(2) There are also two considerations for the formation condition of single $CeAlO_3$ in steel at 1873 K. One is that $MgO \cdot Al_2O_3$ can transform into stable $CeAlO_3$ (as shown in Equation (8)), which means $CeAlO_3$ cannot spontaneously transform into Ce_2O_2S (as shown in Equation (10)); this will provide both the upper and lower limits of Ce activity. The other consideration is that $CeAlO_3$ can directly form from steel (as shown in Equation (3)), but Ce_2O_2S should not spontaneously precipitate (as shown in Equation (4)). This will also limit the upper and lower limits of oxygen activity, which means the oxygen activity should be higher than the equilibrium oxygen activity limited by Ce-Al-O but lower than the equilibrium oxygen activity limited by Ce-O-S, as follows.

$$0.0464 \cdot a_O \cdot a_{Al} \cdot a_S^{-1} > a_{Ce} > 2.061 \times 10^{-4} \cdot a_{Mg}^{-3/4} \cdot a_{Al}^{-1/2},$$
$$6.1 \times 10^{-11} \cdot a_{Ce}^{-1} \cdot a_S^{-1/2} > a_O > 4.31 \times 10^{-7} \cdot a_{Al}^{-1/3} \cdot a_{Ce}^{-1/3}. \tag{14}$$

(3) There are also two considerations for the formation condition of $CeAlO_3$ and Ce_2O_2S in steel at 1873 K. One is that $MgO \cdot Al_2O_3$ can transform into stable $CeAlO_3$ (as shown in Equation (8)), and $CeAlO_3$ can transform into Ce_2O_2S (as shown in Equation (10)); this will provide both the lower limits of Ce activity. The other consideration is that $CeAlO_3$ and Ce_2O_2S can directly form from steel (as shown in Equations (3) and (4)). This will also limit the lower limits of oxygen activity, which means the oxygen activity should be higher than the equilibrium oxygen activity limited by Ce-Al-O and Ce-O-S, as follows.

$$a_{Ce} > \max\left\{2.061 \times 10^{-4} \cdot a_{Mg}^{3/4} \cdot a_{Al}^{1/2}, 3.09 \times 10^{-3} \cdot a_{Mg}^{3/8} \cdot a_{Al}^{3/4} \cdot a_O^{1/2} \cdot a_S^{-1/2}\right\},$$
$$a_O > \max\left\{4.31 \times 10^{-7} \cdot a_{Al}^{-1/3} \cdot a_{Ce}^{-1/3}, 6.1 \times 10^{-11} \cdot a_{Ce}^{-1} \cdot a_S^{-1/2}\right\}. \tag{15}$$

(4) There are also two considerations for the formation condition of single Ce_2O_2S in steel at 1873 K. One is that $MgO \cdot Al_2O_3$ can transform into stable Ce_2O_2S (as shown in Equation (10)), which means $CeAlO_3$ cannot spontaneously transform into $CeAlO_3$ (as shown in Equation (8)); this will provide both the upper and lower limits of Ce activity. The other consideration is that Ce_2O_2S can directly form from steel (as shown in Equation (4)), but $CeAlO_3$ should not spontaneously precipitate (as shown in Equation (3)). This will also limit the upper and lower limits of oxygen activity, which means the oxygen activity should be higher than the equilibrium oxygen activity limited by Ce-O-S but lower than the equilibrium oxygen activity limited by Ce-Al-O, as follows.

$$a_{Ce} > 3.09 \times 10^{-3} \cdot a_{Mg}^{3/8} \cdot a_{Al}^{3/4} \cdot a_O^{1/2} \cdot a_S^{-1/2}, \quad 4.31 \times 10^{-7} \cdot a_{Al}^{-1/3} \cdot a_{Ce}^{-1/3} > a_O > 6.1 \times 10^{-11} \cdot a_{Ce}^{-1} \cdot a_S^{-1/2}. \tag{16}$$

In summary, the modification mechanism of Ce to spinel inclusions can be shown in Figure 6. Ce can modify the spinel inclusions into $CeAlO_3$ or Ce_2O_2S. There are three main modification paths: ① Spinel can directly transform into $CeAlO_3$ by Ce, namely spinel → Ce-Al-O + Mg-Al-O → $CeAlO_3$, which can be proved by the complex inclusion shown in Figure 3a. ② After Ce treatment, spinel can be preferentially transformed to $CeAlO_3$. After a while, the resulting $CeAlO_3$ product layer will be transformed to Ce_2O_2S. The first-step reaction was considered to be difficult to perform completely because we observed some inclusions with a three-layer structure in the experiment, as shown in Figure 3c. If treatment time is enough, it will eventually transform into uniform Ce_2O_2S. The above path can be summarized as spinel → Ce-Al-O + Mg-Al-O → Ce-O-S + Ce-Al-O + Mg-Al-O → Ce_2O_2S. ③ Spinel can be directly transformed into Ce_2O_2S by Ce, the modification path is spinel → Ce-O-S + Mg-Al-O → Ce_2O_2S. The related modification mechanism is as follows: After Ce addition, Mn and Mg in the spinel inclusions are first replaced by Ce, and the spinel structure will be destroyed in the above process to form $CeAlO_3$. If the O content in the steel is low, which means that the S content is relatively high, S in the steel will participate in the above displacement reaction to replace the O in the inclusion. In general, $CeAlO_3$ and spinel inclusions will transform into Ce_2O_2S through paths 2 and 3. Under the experimental steel conditions, the modification sequence of Ce to spinel inclusions is spinel → $CeAlO_3$ → Ce_2O_2S.

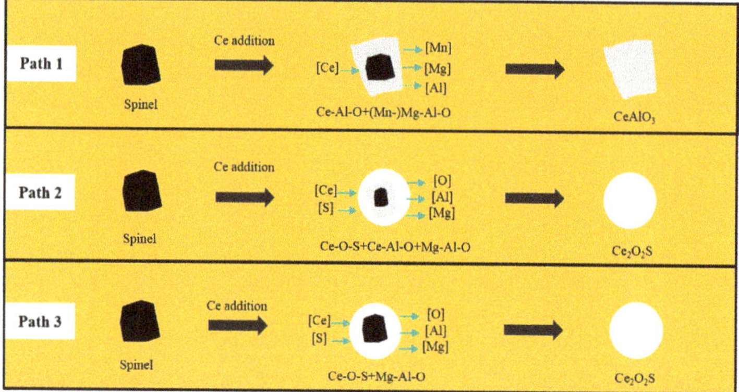

Figure 6. Modification mechanism of spinel inclusions.

Huang et al. [7,9] and Wang et al. [8] reported that the sequence of Ce-modified spinel is spinel $\rightarrow$ CeAlO$_3$ $\rightarrow$ Ce$_2$O$_3$ $\rightarrow$ Ce$_2$O$_2$S or spinel $\rightarrow$ Ce$_2$O$_3$ $\rightarrow$ Ce$_2$O$_2$S. However, no Ce$_2$O$_3$ inclusions were found in the present experimental steel. After comparing the experimental conditions, it was found that the contents of Al and S in the steel were different. Therefore, the stability diagram in the Fe-10%Mn-0.001%Mg-0.015%Ce-0.005%O-Al-S system at 1873 K was calculated using FactSage 7.1 (databases FactPS, FToxid, FSstel, MYDT), as shown in Figure 7. As can be seen from the red line in Figure 7, as the S content in the steel increased, the precipitation condition for the Ce$_2$O$_3$ required a lower Al content. The S content in our experimental steels was as high as 0.015%. Therefore, the formation condition of Ce$_2$O$_3$ was not reached when the Al content was higher than 0.02%.

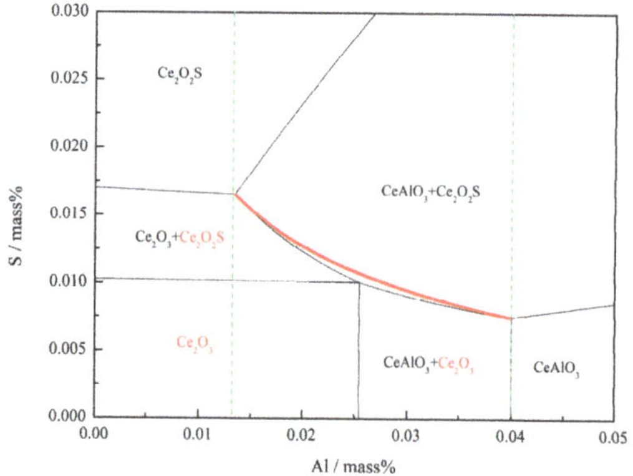

Figure 7. Stability diagram in the Fe-10%Mn-0.001%Mg-0.015%Ce-0.005%O-Al-S system at 1873 K.

5. Ce Content Required for Complete Modification of Spinel Inclusions

There are complex interactions between different elements, such as Ce, Al, O, and S, in the steel during the modification process. Under present experimental conditions, spinel is preferentially transformed to CeAlO$_3$. In order to obtain the minimum Ce content required for complete modification of spinel inclusions, the effect of aluminum and oxygen content on the related Ce content in steel was calculated by FactSage. The content of each element in the steel discussed in this section is total content, and the calculation results are shown in Figure 8.

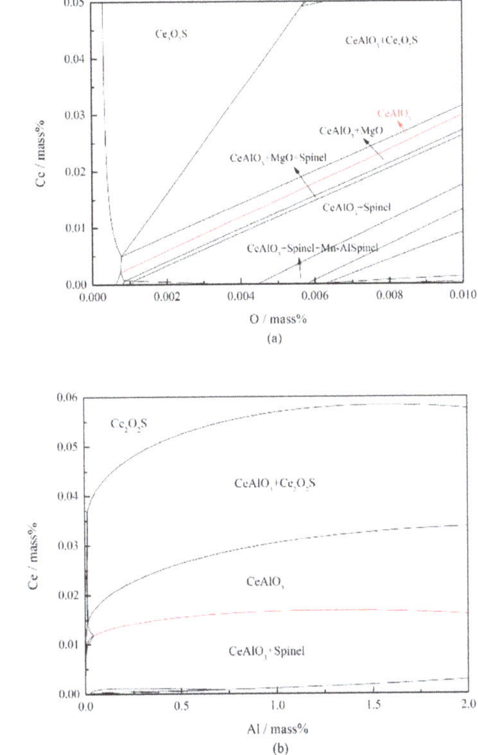

Figure 8. Effect of oxygen and aluminum content on inclusions in steel, (**a**) $w_S = 0.015\%$ and $w_{Al} = 0.02\%$, (**b**) $w_O = 0.0045\%$ and $w_S = 0.015\%$.

Figure 8a shows the effect of oxygen content on inclusions in steel, where $w_{Al} = 0.02\%$ and $w_S = 0.015\%$. It can be seen from the red line in the figure that the minimum Ce content required for modification of spinel inclusions is linear with the oxygen content in the steel. When $w_{Ce} = 2.99 \cdot w_O$, the inclusions in the steel can be controlled as $CeAlO_3$. Wang et al. [8] also obtained a similar conclusion: that increasing the O content would increase the Ce content required for complete modification of spinel inclusions.

Theoretically, the inclusions can be controlled to be $CeAlO_3$ when the Ce content in the steel is three times that of the oxygen content. Therefore, by measuring the oxygen content in the steel, a suitable rare earth Ce addition amount required for complete modification of spinel inclusions can be obtained.

Figure 8b shows the effect of aluminum content on types of stable inclusions in steel, where $w_O = 0.0045\%$ and $w_S = 0.015\%$. It can be seen from the red line in the figure that as the aluminum content in the steel increases, the minimum Ce content required for modification spinel inclusions does not change much, which indicates that the aluminum content has little effect on the minimum Ce content required for modification of spinel inclusions.

6. Conclusions

In present paper, the transformation process of inclusions in medium manganese steel was studied by high-temperature simulation experiments. The modification mechanism of inclusions in steel under different aluminum contents was discussed by thermodynamic calculations. Under the present experimental conditions, the following conclusions were obtained.

(1) Ce treatment can modify spinel inclusions into rare earth inclusions. After Ce treatment, the amount of inclusions in the steel will decrease, and the size of inclusions is basically less than 5 μm.

(2) The modification mechanism of spinel inclusions by Ce is as follows: After Ce is added to steel, Mn and Mg in the spinel inclusions are first replaced by Ce to form $CeAlO_3$, which will cause the destruction of the spinel structure. When the O content in the steel is at a lower level, S in the steel will take part in the replacement reaction to replace the O in the inclusion, then the $CeAlO_3$ and spinel inclusions will be finally transformed into Ce_2O_2S.

(3) In the Fe-Mn-Al-Mg-Ce-O-S system at 1873 K, the conditions for forming single $CeAlO_3$ are $0.0464 \cdot a_O \cdot a_{Al} \cdot a_S^{-1} > a_{Ce} > 2.061 \times 10^{-4} \cdot a_{Mg}^{3/4} \cdot a_{Al}^{1/2}$, $6.1 \times 10^{-11} \cdot a_{Ce}^{-1} \cdot a_S^{-1/2} > a_O > 4.31 \times 10^{-7} \cdot a_{Al}^{-1/3} \cdot a_{Ce}^{-1/3}$; the conditions for simultaneous formation of $CeAlO_3$ and Ce_2O_2S are $a_{Ce} > \max\left\{2.061 \times 10^{-4} \cdot a_{Mg}^{3/4} \cdot a_{Al}^{1/2}, 3.09 \times 10^{-3} \cdot a_{Mg}^{3/8} \cdot a_{Al}^{3/4} \cdot a_{O}^{1/2} \cdot a_S^{-1/2}\right\}$, $a_O > \max\left\{4.31 \times 10^{-7} \cdot a_{Al}^{-1/3} \cdot a_{Ce}^{-1/3}, 6.1 \times 10^{-11} \cdot a_{Ce}^{-1} \cdot a_S^{-1/2}\right\}$; and the conditions for forming single Ce_2O_2S are $a_{Ce} > 3.09 \times 10^{-3} \cdot a_{Mg}^{3/8} \cdot a_{Al}^{3/4} \cdot a_O^{1/2} \cdot a_S^{-1/2}$, $4.31 \times 10^{-7} \cdot a_{Al}^{-1/3} \cdot a_{Ce}^{-1/3} > a_O > 6.1 \times 10^{-11} \cdot a_{Ce}^{-1} \cdot a_S^{-1/2}$.

(4) The minimum Ce content required for complete modification of spinel inclusions is mainly related to the O content. When the total Ce content is three times that of the total O content in the steel, the spinel inclusions can be transformed into $CeAlO_3$.

Author Contributions: Z.X. and C.L. conceived and designed the experiments, analyzed the data, and conducted writing–original draft preparation, editing, and visualization.

Funding: This work was financially supported by the National Key R&D Program of China (No. 2017YFC0805100), the National Natural Science Foundation of China (No. 51874082), and the Fundamental Research Funds for the Central Universities (Grant No. N180725008).

Acknowledgments: Thanks to Jiyu Qiu for providing language help.

References

1. Kong, L.Z.; Deng, Z.Y.; Zhu, M.Y. Formation and evolution of non-metallic inclusions in medium Mn steel during secondary refining process. *ISIJ Int.* **2017**, *57*, 1537–1545. [CrossRef]

2. Shao, X.J.; Ji, C.X.; Cui, Y.; Li, H.B.; Zhu, G.S.; Chen, B.; Liu, B.S.; Bao, C.L. Brief instruction of the effect of inclusions on fatigue behavior of steel. In Proceedings of the National Steelmaking Conference, Xi'an, China, 21 May 2014.

3. Wang, L.; Zhang, P.J.; Lu, J.X. Quality and production technology of steel sheets for automobile. *Iron Steel* **1999**, *34*, 908–912. (In Chinese)

4. Yang, S.F.; Li, J.S.; Wang, Z.F.; Li, J.; Lin, L. Modification of $MgO \cdot Al_2O_3$ spinel inclusions in Al-killed steel by Ca-treatment. *Int. J. Miner. Metall. Mater.* **2011**, *18*, 18–23. [CrossRef]

5. Pretorius, E.B.; Oltmann, H.G.; Cash, T. The effective modification of spinel inclusions by Ca treatment in LCAK steel. *Iron Steel Technol.* **2010**, *7*, 31–44.

6. Ma, W.J.; Bao, Y.P.; Wang, M.; Zhao, L.H. Effect of Mg and Ca treatment on behavior and particle size of inclusions in bearing steels. *ISIJ Int.* **2014**, *54*, 536–542. [CrossRef]

7. Huang, Y.; Cheng, G.G.; Xie, Y. Modification mechanism of cerium on the inclusions in drill steel. *Acta Metall.* **2018**, *54*, 1253–1261.

8. Wang, L.J.; Liu, Y.Q.; Chou, K.C. Evolution mechanisms of $MgO \cdot Al_2O_3$ inclusions by cerium in spring steel used in fasteners of high-speed railway. *Trans. Iron Steel Inst. Jpn.* **2015**, *55*, 970–975. [CrossRef]

9. Huang, Y.; Cheng, G.G.; Li, S.J.; Dai, W.X. Effect of Cerium on the Behavior of Inclusions in H13 Steel. *Steel Res. Int.* **2018**, *89*, 1800371. [CrossRef]

10. Hirata, H.; Isobe, K. Steel having finely dispersed inclusions. U.S. Patent 20060157162A1, 20 July 2006.

11. Adabavazeh, Z.; Hwang, W.S.; Su, Y.H. Effect of adding cerium on microstructure and morphology of Ce-based inclusions formed in low-carbon steel. *Sci. Rep.* **2017**, *7*, 1–10. [CrossRef] [PubMed]

12. Wilson, W.G.; Kay, D.A.R.; Vahed, A. The use of thermodynamics and phase equilibria to predict the behavior of the rare earth elements in steel. *JOM* **1974**, *26*, 14–23. [CrossRef]

13. Fruehan, R.J. The free energy of formation of Ce_2O_2S and the nonstoichiometry of cerium oxides. *Metall. Mater. Trans. B* **1979**, *10*, 143–148. [CrossRef]

14. Paek, M.K.; Jang, J.M.; Kang, Y.B.; Pak, J.J. Aluminum deoxidation equilibria in liquid iron: Part I. experimental. *Metall. Mater. Trans. B* **2015**, *46*, 1826–1836. [CrossRef]

15. Vahed, A.; Kay, D.A.R. Thermodynamics of rare earths in steelmaking. *Metall. Mater. Trans. B* **1976**, *7*, 375–383. [CrossRef]

16. Du, T. Thermodynamics of rare earth elements in iron-base solutions. *J. Iron. Steel Res.* **1994**, *6*, 6–12.

17. Hong, T.; Debroy, T. Time-temperature-transformation diagrams for the growth and dissolution of inclusions in liquid steels. *Scr. Mater.* **2001**, *44*, 847–852. [CrossRef]

18. Itoh, H.; Hino, M.; Ban, Y.S. Thermodynamics on the formation of non-metallic inclusion of spinel ($MgO \cdot Al_2O_3$) in liquid steel. *Tetsu-to-Hagané* **1998**, *84*, 85–90. [CrossRef]

19. Rohde, L.E.; Choudhury, A.; Wahlster, M. Neuere Untersuchungen über das Aluminium-Sauerstoff-Gleichgewicht in Eisenschmelzen. *Archiv für das Eisenhüttenwesen* **1971**, *42*, 165–174. [CrossRef]

20. Seo, J.D.; Kim, S.H.; Lee, K.R. Thermodynamic assessment of the Al deoxidation reaction in liquid iron. *Steel Res. Int.* **1998**, *69*, 49–53. [CrossRef]

MDPI

St. Alban-Anlage 66

4052 Basel

Switzerland

Tel. +41 61 683 77 34

Fax +41 61 302 89 18

www.mdpi.com

Metals Editorial Office

E-mail: metals@mdpi.com

www.mdpi.com/journal/metals